JOSEPH F MICELI JR

The Essential Guide to IAM 3.0 Success

Adapt, Defend, Evolve.

CYBICUS
DIGITAL

Contents

Introduction

We are living through a fundamental shift in the landscape of digital trust. The very boundaries of what we once understood as Identity and Access Management (IAM) are being reshaped,sometimes visibly, often invisibly,by forces like artificial intelligence, machine learning, and quantum computing. Enterprises now sit at the crossroads of innovation and risk, where the tools used to enable incredible new efficiencies are also producing unprecedented vulnerabilities.

The old IAM world was defined by static controls and deterministic rules: you were who your credentials said you were, and access was defined by your role or your group. But this binary paradigm no longer holds its ground. Every day brings new reports of cyberattacks that exploit gaps in legacy authentication, or harness the relentless speed and cunning of AI-driven adversaries. Bad actors are no longer just hackers in hoodies, they are automated, self-learning systems built to probe and defeat our very notions of identity.

Against this backdrop emerges IAM 3.0, an ecosystem where everything is dynamic, contextual, and subject to machine reasoning. Access is now determined by a constellation of signals: behavioral patterns, device finger-prints, peer contexts, and real-time risk scores. AI and ML promise to reduce workload and surface hidden threats,yet the very intelligence we embed also carries the seeds of new forms of subversion. Data can be poisoned. Models can drift. Autonomy can become an attack vector. Meanwhile, the rising tide of quantum computing threatens our cryptographic bedrock, making even the strongest encryption vulnerable to tomorrow's compute breakthroughs.

The journey to IAM 3.0 is about more than just technological evolution; it is a reckoning with the true nature of trust in an era shaped by machine intelligence. We must ask,when decisions move at the speed of algorithms, how do we hold them accountable? When identity is governed not by credentials, but by behaviors and context, how do we ensure governance, transparency, and fairness? And as decentralized identity and sovereign AI carve up the digital world into new borders, how do we maintain unity,of policy, of purpose, of security?

This book is your field guide for navigating this volatile, high-stakes landscape. Across its chapters, we'll unpack the architectural choices facing IAM leaders today, illuminate the dark corners exploited by AI-enabled adversaries, and map out strategies for embedding resilience in the face of quantum disruption. We'll explore not only the dangers, but the immense opportunities presented by intelligent, adaptive IAM,if we build with care, foresight, and a relentless focus on explainability and ethics.

IAM 3.0 is not just an upgrade of old platforms; it is a wholesale reimagining of how we create, evaluate, and broker trust in the digital age. As identity becomes both our gatekeeper and our battleground, the stakes could not be higher. The time to adapt is now. This guidebook is your companion for leading the charge into the emerging future,one where access is smart, risk is contextual, and trust is the most valuable asset of all.

* * *

1

The Rise of IAM 3.0: Breaking Old Paradigms

For decades, the world of Identity and Access Management was a predictable, if somewhat rigid, place. It was a kingdom ruled by directories, defined by perimeters, and governed by a simple, comforting mantra: a user's role dictates their access. This was the era of IAM 1.0 and its successor, 2.0, a time when the digital world, for all its growth, still fit into relatively neat boxes. We built digital walls around our networks, erected castles of on-premises servers, and gave our users keys in the form of passwords and group memberships. The model was straightforward because the landscape was, too.

In the earliest days of IAM 1.0, identity was synonymous with a user account sitting in a corporate directory, like Novell eDirectory or Microsoft Active Directory. The primary concern was managing access to internal file shares, printers, and monolithic applications running within the four walls of the office. Security was about the perimeter; as long as you were inside the trusted network, you were generally considered safe. The "castle-and-moat" architecture wasn't just a metaphor; it was the accepted blueprint for enterprise security. The drawbridge was the firewall, and the guards were simple username and password prompts.

As the web exploded, this model evolved into IAM 2.0. The perimeter began to stretch. Web Access Management (WAM) gateways were created to extend corporate authentication to browser-based applications. We saw the rise of federation protocols like SAML, which allowed identities to be asserted across organizational boundaries, giving birth to Single Sign-On (SSO). For the first time, a user's corporate identity could be used to log into a third-party service like Salesforce without creating a new account. It was a significant leap forward, making access more convenient and centralized.

The core logic, however, remained largely the same. It was deterministic. Access was granted based on a set of predefined, static rules. The dominant paradigm was Role-Based Access Control (RBAC), a system where permissions were bundled into roles corresponding to job functions. An employee in the accounting department was assigned the "Accountant" role and, like magic, received access to the financial systems. A new sales representative got the "Sales" role and could instantly access the CRM. It was orderly, auditable, and for a time, it was enough.

But the digital world refused to stay in its neat boxes. The ground began to shift, slowly at first, then with tectonic force. The very foundations upon which IAM 1.0 and 2.0 were built started to crack. The concept of a defensible perimeter, the central pillar of the old security model, began to dissolve into irrelevance. The castle walls were breached not by a battering ram, but by a thousand different entry points opening up all at once. The moat evaporated under the heat of technological progress.

The first major force was the cloud. It wasn't just one cloud, but a sprawling, multi-vendor universe of Infrastructure-as-a-Service (IaaS), Platform-as-a-Service (PaaS), and Software-as-a-Service (SaaS). Suddenly, critical applications and sensitive data no longer resided within the corporate data center. They were scattered across servers owned by Amazon, Microsoft, and Google, and in applications run by countless other vendors. The on-premises Active Directory, once the single source of truth for identity, was now just one voice in a chorus.

Trying to manage access in this hybrid, multi-cloud world with legacy IAM tools became a nightmare of complexity. Each cloud platform had its own identity system, its own way of defining roles and permissions. Stitching them all together with on-premises controls was like trying to patch a quilt with duct tape. Synchronization was brittle, visibility was fragmented, and the old WAM gateways were not designed to protect access to the control planes of an entire cloud infrastructure. The perimeter was no longer a line on a network diagram; it was a dotted, ever-shifting boundary encompassing the entire internet.

Next came the API economy. In the old model, access was primarily about humans logging into applications. But modern software is not built as a monolith; it's a constellation of microservices, all communicating with each other through Application Programming Interfaces (APIs). This created an explosion of non-human identities. Scripts, services, applications, and containers all needed to authenticate and be authorized to access data and other services. These machine identities vastly outnumbered human users, operated 24/7, and didn't fit neatly into the RBAC model.

You can't assign a "job title" to a microservice. You can't ask a container to go through an access certification campaign to attest to its permissions. The deterministic, role-based logic of IAM 2.0 buckled under the sheer volume and dynamic nature of machine identities. Attackers quickly realized this was a new, fertile ground for exploitation. A compromised API key could grant silent, persistent access to backend systems, completely bypassing security controls focused on human users.

The final blow to the old paradigm was the decentralization of the work-force. The shift to remote and hybrid work, accelerated by global events, permanently dismantled the notion of a "trusted" internal network. When employees connect from their homes, from coffee shops, and from airports, using a mix of corporate and personal devices, the concept of "inside the perimeter" becomes meaningless. Every access request, regardless of its origin, is now effectively an external request. The network location of a user is no longer a reliable indicator of trust.

This confluence of cloud, APIs, and remote work rendered the castle-and-moat model obsolete. Security professionals came to a sobering conclusion: the perimeter was gone. In its place, a new organizing principle for security had to be found. If the network could no longer be trusted, and devices could no longer be implicitly trusted, the only constant in any access transaction was the identity of the entity,human or machine,making the request. Identity had become the new perimeter.

This realization is the genesis of IAM 3.0. It is not an incremental update or a new product category. It represents a fundamental rewiring of how we think about access. It is a paradigm shift from the static, deterministic world of its predecessors to a dynamic, probabilistic, and context-aware model. IAM 3.0 starts with the assumption that the network is hostile and that any access request could be malicious until proven otherwise. It's a philosophy built on a simple but powerful principle: never trust, always verify.

This principle, widely known as Zero Trust, is the beating heart of IAM 3.0. It demands that trust is never granted implicitly but must be continuously earned and re-evaluated. An employee who authenticated successfully from their corporate laptop in the morning is not automatically trusted in the afternoon. Their context may have changed. Are they suddenly accessing a sensitive system they've never touched before? Is their request originating from an unusual geographic location? Is their device showing signs of compromise?

In the IAM 3.0 world, access decisions are no longer a simple "yes" or "no" based on a static role. They are the output of a real-time risk calculation that ingests dozens of signals. The identity of the user is just one data point. Others include the health and posture of the device, the location of the request, the time of day, the sensitivity of the resource being accessed, and the user's normal behavior. Access becomes a fluid continuum, where the level of assurance required changes dynamically based on the risk of the transaction.

This shift necessitates a move away from the one-and-done authentication event at the beginning of a session. Instead, IAM 3.0 advocates for continuous authentication and authorization. Trust must be validated not just at the point of login, but throughout the user's session. A low-risk activity, like reading a company-wide announcement, might require no extra checks. But attempting to download a sensitive customer database should trigger a step-up authentication challenge, such as a biometric prompt, to re-verify the user's identity in that specific, high-risk context.

The technologies that underpin this new paradigm are a mix of mature standards and emerging innovations. Federation protocols, the stars of the IAM 2.0 era, remain critically important. Standards like SAML, OAuth 2.0, and OpenID Connect (OIDC) provide the essential plumbing needed to create a unified identity fabric that connects users to the vast ecosystem of cloud and SaaS applications. They are the lingua franca of identity on the modern internet, enabling identity information to be securely exchanged between different systems and platforms.

However, the way these protocols are used is evolving. They are no longer just for enabling SSO for human users. They are now central to securing API access (via OAuth 2.0) and creating standardized identity layers for a wide range of applications. They provide the transport mechanism for the rich contextual signals that IAM 3.0 systems rely on to make intelligent access decisions.

At the same time, IAM 3.0 is aggressively breaking the paradigm of the password. For decades, passwords have been the weakest link in the security chain,easily stolen, guessed, or phished. The move to passwordless authentication is a cornerstone of the new model. Technologies like the FIDO2 standards, which leverage public-key cryptography through device-bound authenticators like Windows Hello, Apple's Face ID/Touch ID, or physical security keys, are fundamentally changing the game.

By using biometrics or a simple user gesture to unlock a cryptographic private key stored securely on the device, FIDO2 eliminates shareable secrets. There is no password for an attacker to steal from a server database or for a user to be tricked into revealing. This single change short-circuits a huge percentage of common attack vectors, from phishing to credential stuffing. It breaks the old model of security being reliant on something the user knows and shifts it to something they have (the device) and something they are (their biometric).

This focus on stronger, more user-friendly authentication is part of a broader acknowledgment that security cannot come at the expense of productivity. If security controls are too cumbersome, users will find ways to bypass them. The elegance of passwordless methods is that they are often both more secure and less disruptive than typing a complex password. This balance between security and user experience is a key design principle of IAM 3.0.

The new paradigm also forces a reckoning with the explosive growth of non-human identities. Machines, services, bots, and APIs can no longer be an afterthought. IAM 3.0 treats machine identity management as a first-class discipline, on par with human identity governance. This involves discovering and inventorying all non-human identities, issuing and managing their credentials (like API keys and TLS certificates), and enforcing policies of least privilege. A service should only have the exact permissions it needs to perform its function, for the shortest possible duration.

This is a far cry from the old practice of embedding credentials in configuration files or creating a single, overly-privileged "service account" shared by multiple applications. Modern approaches involve dynamic secret management, where credentials are created just-in-time and automatically rotated, dramatically reducing the window of opportunity for an attacker if a secret is compromised.

Emerging on the horizon is an even more radical break with the past: the concept of decentralized identity, often associated with Self-Sovereign Identity (SSI). This model challenges the fundamental paradigm of identity being issued and controlled by a central authority, whether that's a corporation or a government. Instead, it puts the individual in control of their own identity attributes, stored in a digital wallet on their own device. They can then present verifiable, cryptographically signed credentials to prove specific facts about themselves without revealing unnecessary information.

For example, to prove they are over 21, a user could present a verifiable credential from their digital wallet without also having to reveal their name, address, or exact date of birth. This is a profound shift toward privacy and user consent, breaking the old model where identity providers became massive repositories of personal data. While still in its early stages, decentralized identity represents a potential future pillar of IAM 3.0, further distributing trust away from centralized silos.

For architects, CISOs, and IT leaders, navigating this transition requires more than just buying new tools. It demands a fundamental change in mindset. The objective is no longer simply to provision and deprovision user accounts. The new mission is to manage access risk dynamically, across the entire digital ecosystem. This means shifting from a compliance-driven, checklist-based approach to a risk-driven, intelligence-based one.

The traditional quarterly or annual access certification, where a manager scrolls through a long list of their employees' entitlements and clicks "approve," is an artifact of a slower, more static world. In the dynamic environment of IAM 3.0, access rights can change from moment to moment. Governance must therefore become continuous and event-driven. An alert should be triggered not three months after the fact, but in the instant that a user's access pattern deviates from the norm or a risky combination of entitlements is created.

This also means that IAM can no longer be viewed as a siloed IT cost center. It is a strategic enabler of business agility and a core pillar of the organization's security posture. A well-designed IAM 3.0 architecture allows the business to adopt new cloud services securely, onboard partners and contractors with minimal friction, and empower employees to work from anywhere, on any device. It transforms security from a department of "no" to a function that enables the business to move faster, more safely.

Ultimately, the goal has been reframed. The old mission was to keep the bad guys out. The new mission is far more nuanced: to enable the right entity (human or machine) to access the right resource, at the right time, from the right device, and under the right conditions, all with the least possible friction. It is a delicate balancing act between security, productivity, and user experience. This is the promise, and the challenge, of breaking the old paradigms and embracing the rise of IAM 3.0.

* * *

2

The Threat Landscape: AI, ML, and Quantum Adversaries

The image of the adversary that many security professionals grew up with is now a relic, a comfortable but dangerously outdated stereotype. We picture a lone figure, perhaps a disaffected youth or a rogue agent, hunched over a keyboard in a dimly lit room, manually probing firewalls and crafting clever exploits. This human-centric view of the attacker, while once largely accurate, has been rendered obsolete. The modern threat landscape is not populated by individuals engaged in artisanal hacking; it is an industrialized ecosystem powered by automation, orchestrated by intelligent systems, and already preparing for a future where today's encryption is rendered meaningless.

The adversary we face today is less a single person and more a distributed, self-learning entity. It is a nation-state's cyberwarfare division with the budget of a small country, a profit-driven ransomware syndicate operating with the cold efficiency of a Fortune 500 company, and an ever-growing network of semi-autonomous bots that can execute attacks at a scale and speed that defies human comprehension. They are not just using better tools; they are weaponizing the very technologies we are pinning our hopes on for a more secure future: Artificial Intelligence and Machine Learning. And they are patiently waiting for the dawn of a quantum world that will give them the ultimate skeleton key.

Understanding this new adversary is the first, and most critical, step in building a resilient IAM 3.0 strategy. Our defenses cannot be designed to stop the ghost of attackers past; they must be architected to withstand the intelligent, adaptive, and relentless threats of today and tomorrow. The battle has changed, and to fight it, we must first know our enemy.

The most immediate and palpable shift in the threat landscape is the widespread adoption of artificial intelligence and machine learning by our opponents. This isn't a futuristic concept; it's happening right now, every day. Attackers are using AI not just to make their old methods faster, but to invent entirely new classes of attack that were previously impossible. This has transformed their capabilities across the entire lifecycle of an attack, from initial reconnaissance to final exfiltration.

At the most basic level, AI provides the fuel for hyper-automation. Where a human attacker might scan a few dozen potential targets for vulnerabilities, an AI-driven system can scan millions of IP addresses, websites, and open-source code repositories simultaneously. These are not simple port scans; they are intelligent probes that can identify complex misconfigurations in cloud environments, discover exposed API endpoints, and find subtle vulnerabilities in application logic. The attacker can effectively "ask" their AI to find every unpatched instance of a specific software version across an entire industry sector and have the results in minutes.

This automation extends to the attack itself. Once a vulnerability is identified, an AI agent can be dispatched to exploit it, move laterally within the network, and locate high-value data. This process, which might have taken a human attacker weeks of careful, manual work, can now be compressed into hours or even minutes. The speed of the attack outpaces the ability of most human-led security operations centers (SOCs) to detect and respond. By the time an analyst sees the first alert, the automated attacker may have already achieved its objectives.

Perhaps more insidiously, attackers are leveraging generative AI to perfect the art of deception. The era of poorly worded phishing emails with suspicious links is drawing to a close. Modern adversaries use generative AI models, similar to the ones powering popular chatbots, to create flawless, contextually aware, and highly personalized social engineering lures. These systems can scrape a target's LinkedIn profile, company website, and recent public activity to craft an email that appears to come from a trusted colleague or a legitimate business partner, discussing a relevant project or a recent event.

The sophistication is staggering. An AI can generate a spear-phishing email to a finance employee that references a specific, recent invoice number, or craft a message to an IT administrator that credibly discusses an ongoing system migration. This level of personalization at scale was once the exclusive domain of highly resourced intelligence agencies. Now, it is becoming commoditized, available to any criminal group willing to pay for access to the right tools.

This AI-driven deception is evolving from text to sight and sound. The rise of deepfake technology represents a new frontier for identity fraud. Attackers can now use AI to generate realistic video and audio of executives, tricking employees into making unauthorized wire transfers or divulging sensitive information. Imagine receiving a video call from your CEO, their face and voice perfectly replicated, asking you to urgently process a payment. How many would hesitate? These "vishing" (voice phishing) and deepfake video attacks directly target the human element of security, bypassing technical controls by manipulating our innate trust in what we see and hear. They pose a direct challenge to identity verification processes that rely on video calls or voice biometrics.

Beyond automating and personalizing attacks, malicious AI is being designed for stealth and adaptation. Attackers are developing malware that uses machine learning to change its own behavior to evade detection. This "polymorphic" or "metamorphic" malware can rewrite its own code, alter its network communication patterns, or pause its activity when it detects the presence of a sandbox or analysis tool. It learns what "normal" looks like inside a victim's network and then mimics that behavior, making it incredibly difficult for anomaly detection systems to flag its activity as malicious.

We are entering an era of adversarial cat-and-mouse, where our defensive AI models are pitted directly against the attackers' offensive AI. The adversary's system learns to generate attack patterns that fall just below the detection threshold of our security tools. It probes our defenses, learns the rules, and then devises a strategy to circumvent them. This is a far cry from fighting against a static list of known virus signatures or IP blocklists; it's like trying to catch a shapeshifter.

As we deploy more sophisticated AI and machine learning models to defend our enterprises, we must face a chilling reality: these very systems are themselves becoming prime targets. Attackers are no longer content with just bypassing our AI-driven defenses; they are actively seeking to corrupt them, turning our own intelligence against us. This field, known as adversarial machine learning, represents a fundamental threat to the probabilistic decision-making at the heart of IAM 3.0.

One of the most concerning techniques is AI poisoning, or data poisoning. Machine learning models learn by being trained on vast datasets. An AI model designed to detect fraudulent access requests, for instance, is fed millions of examples of both legitimate and fraudulent logins. Over time, it learns to recognize the subtle patterns that distinguish one from the other. But what if an attacker could subtly manipulate that training data?

By injecting a small amount of carefully crafted malicious data into the training set, an attacker can create a hidden backdoor in the AI model. They could, for example, "teach" the model that access requests originating from a specific, attacker-controlled IP address are always legitimate. Once the poisoned model is deployed, the attacker can waltz right through the front door, their activity flagged as perfectly normal by the very system designed to stop them. The corruption is silent, buried within the millions of weighted connections inside the neural network, making it nearly impossible to detect through traditional code reviews.

Another common approach is a model evasion attack. In this scenario, the attacker doesn't tamper with the model's training. Instead, they probe the live, trained model to understand how it makes decisions. They test its boundaries, feeding it slightly modified inputs until they find its blind spots. For example, an attacker might craft a malicious payload that, to a human, is clearly an attack, but due to a quirk in how the AI model processes data, it gets classified as benign. They have effectively found a loophole in the AI's logic, allowing them to slip past its defenses.

These attacks on the machine learning models themselves represent a systemic risk. A single compromised IAM AI model could grant unauthorized access across an entire organization. It undermines the very core of probabilistic trust. When you can no longer trust the risk score generated by your AI, the entire IAM 3.0 model of continuous, behavior-based evaluation begins to crumble. It's a sophisticated attack that doesn't break a door down; it subtly convinces the guard that you belong inside.

Looming behind the immediate threats of AI and ML is a more profound, existential risk: the quantum time bomb. While practical, large-scale quantum computers are still some years away, their eventual arrival threatens to catastrophically undermine the cryptographic foundations upon which the entire digital world is built. This is not a distant, academic problem; the danger has already begun.

Today's most widely used public-key encryption algorithms, such as RSA and Elliptic Curve Cryptography (ECC), rely on the mathematical difficulty of factoring very large numbers. A classical computer, even a supercomputer, would take billions of years to break a standard RSA-2048 key. A sufficiently powerful quantum computer, however, could theoretically break it in a matter of hours or even minutes using algorithms like Shor's algorithm.

This has devastating implications for identity and access management. The security of nearly every authentication and data protection mechanism we use relies on these very algorithms. This includes the digital certificates that verify the identity of websites (TLS/SSL), the SAML and OAuth tokens that enable federated single sign-on, the digital signatures that ensure data integrity, and the security of hardware tokens and smart cards. When the cryptography breaks, all of these systems become vulnerable. An attacker could impersonate any website, forge any authentication token, and decrypt any secure communication.

The most pressing quantum threat is encapsulated in the strategy known as "Harvest Now, Decrypt Later." Highly sophisticated adversaries, particularly nation-states, are well aware of the coming quantum revolution. They are actively engaged today in siphoning off and storing massive amounts of encrypted data from governments, corporations, and infrastructure providers. They cannot break the encryption today, but they are betting that they will be able to in the future, once they possess a quantum computer.

Every piece of encrypted data we transmit today,every authenticated session, every stored password hash, every VPN tunnel's traffic,is potentially being warehoused by an adversary. This turns today's secure data into a ticking time bomb. An encrypted backup of a user credential database, stolen today, might be useless to the attacker now. But in five or ten years, it could become a treasure trove of plaintext passwords and identity information. This long-game strategy fundamentally changes the risk calculus, making the quantum threat a present-day reality, not a future concern.

The profile of the modern adversary is no longer monolithic. We are facing a diverse and specialized ecosystem of attackers, each with different motivations, resources, and techniques, but all increasingly leveraging advanced technology. At the top of the pyramid are the nation-state actors. These groups are patient, exceptionally well-funded, and play a long game. They are the ones most likely to be engaged in "Harvest Now, Decrypt Later" strategies and to be developing bespoke offensive AI capabilities for espionage, sabotage, and information warfare. Their goal is not immediate financial gain, but long-term strategic advantage.

Just below them are the large, organized cybercrime syndicates. These groups operate like multinational corporations, with specialized divisions for malware development, operations, money laundering, and even customer service for their ransomware victims. They are profit-driven, ruthlessly efficient, and are rapid adopters of new technology. They use AI to scale their phishing and ransomware campaigns, identify the most lucrative targets, and automate their attacks to maximize return on investment.

Finally, the proliferation of "as-a-service" models has democratized advanced cybercrime. Less sophisticated actors, from small criminal gangs to individual hacktivists, can now rent or buy access to powerful tools. They can purchase access to a botnet for a distributed denial-of-service (DDoS) attack, use a Phishing-as-a-Service kit to launch a convincing campaign, or leverage Ransomware-as-a-Service platforms where the developer takes a cut of the profits. This lowers the bar for entry, meaning organizations face a threat not just from elite hackers, but from a vast number of less skilled but still dangerous adversaries armed with powerful, off-the-shelf weaponry.

This complex and evolving threat landscape is the crucible in which IAM 3.0 is being forged. The old models of static rules and perimeter-based security were designed for a world of human-speed, predictable attacks. They are fundamentally unequipped to handle adversaries who operate at machine speed, use AI to learn and adapt, and are poised to shatter the cryptographic trust that underpins everything we do.

The necessity of IAM 3.0 is a direct response to this new reality. When the attacker can perfectly mimic a legitimate user's behavior, we need continuous, real-time behavioral analytics to spot the subtle deviations that give them away. When deepfakes can fool a human, we need stronger, cryptographically-bound authentication methods like passwordless FIDO2 that are resistant to such manipulation. When adversaries are poisoning our AI models, we need a new discipline of explainable AI (XAI) and model security to ensure the integrity of our automated decisions. And as the quantum clock ticks, we must urgently begin the migration to post-quantum cryptography to safeguard our identities for the future. The fight for control of our digital world is a fight for control over identity, and the adversary has never been more formidable.

* * *

3

Deterministic vs Probabilistic IAM: A Sea Change

For generations of IT professionals, the logic of access control was as rigid and comforting as the laws of physics. It was a world built on certainty, a digital landscape governed by clear, unwavering rules. This was the kingdom of deterministic Identity and Access Management, a paradigm where every access decision was a foregone conclusion, predictable down to the last bit and byte. If a user possessed the right key, the door would open. It was simple, it was absolute, and for a long time, it was sufficient.

In a deterministic system, the outcome is determined entirely by its initial state and inputs. There is no room for ambiguity, no element of chance. The logic is binary: true or false, yes or no, allow or deny. The most common manifestation of this thinking was, and often still is, Role-Based Access Control (RBAC). The logic is beautifully simple: IF a user is a member of the "Engineers" group, THEN grant access to the source code repository. The system checks the condition, finds it to be true, and the gate swings open.

This model was perfect for the structured, hierarchical organizations of the past. It brought order to the chaos of early user management, allowing administrators to manage permissions for large groups of people efficiently. Instead of assigning permissions one by one to hundreds of users, they could simply manage a handful of roles. When a new engineer was hired, they were dropped into the "Engineers" role and inherited all the necessary access. When they left the company, the role was revoked, and their access vanished. It was clean and, in theory, easy to audit.

As the needs of the enterprise grew more complex, a more granular deterministic model emerged: Attribute-Based Access Control (ABAC). ABAC extended the simple 'if-then' logic of RBAC into a more powerful policy engine. It could make decisions based not just on a user's role, but on a rich set of attributes. A policy might state: IF the user's attribute 'department' is "Finance" AND the resource's attribute 'sensitivity' is "Confidential" AND the current time is between 9 AM and 5 PM, THEN allow access.

This was a significant step forward in sophistication. ABAC allowed for much finer-grained control, enabling policies that could adapt to more context than a simple role ever could. Yet, at its core, it remained fundamentally deterministic. For any given set of attributes, the outcome was still a hardcoded, predetermined "yes" or "no." The system was essentially a very complex flowchart, but a flowchart nonetheless. If you followed the lines, you would always arrive at the same, unchangeable conclusion.

The fatal flaw in this deterministic worldview is its brittleness. These systems are utterly blind to context that hasn't been explicitly coded into a rule. The RBAC system doesn't care if the engineer is logging in at 3 AM from a previously unknown country. The ABAC system, unless specifically programmed with a location attribute, is equally indifferent. It sees that the conditions of its rule have been met and dutifully grants access, oblivious to the blaring red flags surrounding the request.

Deterministic systems have no concept of "unusual." They operate with a dangerous form of certainty, treating every valid request as equally trustworthy. The CEO logging in from her corporate-issued laptop in her office is given the same green light as an attacker who has successfully stolen her credentials and is logging in from a disposable virtual machine hosted in a hostile nation. As long as the credentials are correct and the role membership is valid, the deterministic engine sees no difference. It lacks the faculty of suspicion.

This inherent rigidity is what adversaries have exploited for years. They don't need to break the rules; they just need to follow them using stolen keys. Once they have compromised a legitimate identity, they can operate within the confines of the deterministic system, and as long as they stay within the permissions granted to that identity, the system itself will see nothing wrong. The model is too inflexible to handle the nuance and deception of modern threats. This is the critical juncture where the sea change begins,the shift from the world of certainties to the world of probabilities.

Probabilistic IAM operates on a fundamentally different premise. It accepts that in a complex, hostile digital environment, absolute certainty is an illusion. Instead of asking "Is this user allowed access?", it asks a more intelligent question: "Given everything we know, what is the probability that this access request is legitimate and safe?" The answer is not a simple "yes" or "no," but a confidence score,a numerical representation of risk.

Imagine an access request generating a score between 0 and 1. A score of 0.99 might signify extremely high confidence that the request is legitimate, while a score of 0.2 might indicate a very high probability of malicious intent. This single change,from a binary output to a continuous spectrum of risk,transforms everything. The IAM system is no longer a rigid gatekeeper but an intelligent risk assessment engine.

This engine is fueled by a constant stream of signals, far beyond the static attributes of the old model. It considers the user's historical behavior, creating a baseline of what "normal" looks like for that individual. It assesses the device they are using, checking its health, patch level, and whether it's a known and trusted endpoint. It analyzes the location of the request, calculating the physical possibility of travel since the last known login. It even looks at the request in the context of the user's peer group, is this a resource that people in their role typically access?

Each of these signals contributes to the final risk score. A familiar device and location might push the score up, increasing confidence. An unusual time of day or an attempt to access a highly sensitive resource for the first time might push it down, increasing the measured risk. This is a living, breathing assessment that evaluates trust not as a static state, but as a dynamic variable that changes with every transaction.

Let's revisit our financial controller, Alice, who needs to approve a large wire transfer. This scenario perfectly illustrates the chasm between the two paradigms.

In the deterministic world of IAM 2.0, the process is clinical. The system asks a series of simple questions. Is Alice's username and password correct? 'Yes.' Is Alice a member of the "Financial Controllers" RBAC group? 'Yes.' Has the request been initiated during normal business hours as defined by the ABAC policy? 'Yes.' The flowchart is complete. The system's verdict is absolute: "ALLOW". The transfer is approved. The system is blind to the fact that Alice, who has never worked outside the New York office, appears to be making this request from an internet cafe in a country she's never visited, using a brand new personal device.

Now, let's run the same scenario through a probabilistic IAM 3.0 engine. The process begins similarly, checking the basic credentials. But that's just the opening act. The system immediately begins to calculate a real-time risk score, layering contextual signals on top of the basic authentication.

First, it consults Alice's behavioral baseline. The engine notes that over the past three years, Alice has initiated 1,472 wire transfers, all from her corporate-issued desktop on the 12th floor of their Manhattan headquarters, and always between 9:15 AM and 4:30 PM Eastern Time. Today's request is from a new, unknown device fingerprint and is occurring at what would be 3:00 AM in her home time zone. The model's confidence in the request's legitimacy immediately drops.

Next, the engine performs a geo-velocity check. Alice's last authenticated action was swiping her badge to exit the New York office building just six hours ago. This new request is originating from an IP address in Eastern Europe. The system calculates that it is physically impossible for her to have traveled that distance in that time. This single data point causes the risk score to plummet dramatically. This isn't just unusual; it's a direct contradiction.

The device posture check fails next. The system's endpoint agent reports that the device making the request is a personal laptop, not a corporate-managed one. It's running an outdated operating system and is missing three critical security patches. More risk is added to the calculation. The request is looking less like Alice and more like an attacker who has successfully phished her credentials.

Finally, the engine consults a peer group analysis model. It looks at the behavior of the other 12 financial controllers in the company. In the last five years, none of them have ever initiated a wire transfer of this magnitude from an unmanaged device outside of North America. Alice's request is a radical outlier compared to the collective behavior of her peers.

The probabilistic engine now aggregates these weighted signals. The impossible geo-velocity, the anomalous behavior, the risky device, and the deviation from peer norms all combine to produce a final confidence score of, for example, 0.05,a 95% probability of fraud. The system's response is no longer a simple "allow." Based on a pre-configured policy, a score this low triggers a decisive, multi-pronged response.

The transaction is immediately "BLOCKED". Alice's session is terminated. Her account is automatically suspended to prevent any further malicious activity. A high-priority, context-rich alert is sent to the Security Operations Center, detailing exactly why the risk score was so high: "Impossible travel detected," "Anomalous device and time," "Deviation from peer group behavior." Simultaneously, an automated notification is sent to Alice's manager via a secure channel. The intelligent system didn't just prevent a fraudulent transfer; it initiated the incident response process automatically.

This sea change from deterministic to probabilistic logic has profound implications across the organization. For governance and compliance teams, the world becomes more complicated. An auditor can no longer simply check if a user was in the right group. The audit trail now reads like a narrative of risk assessment. The question isn't "Did the system follow the rule?" but "Was the system's risk calculation reasonable and justifiable?" This creates a powerful incentive for vendors and implementers to ensure their AI models are not opaque black boxes, driving the need for the explainable AI (XAI) that will make these decisions transparent and auditable.

For security analysts, the nature of their work shifts from chasing ghosts to being digital detectives. Instead of sifting through thousands of low-context, binary alerts like "failed login," they receive a smaller number of high-fidelity, risk-based alerts. The alert itself contains the "why",the specific signals that contributed to the high-risk score. This allows them to immediately focus their investigation on the most critical threats, armed with a wealth of information from the moment the incident is created.

Perhaps counter-intuitively, this shift can lead to a dramatically improved user experience. Deterministic systems, in their attempt to be secure, often resort to blunt instruments, forcing every user to jump through the same security hoops for every action. A probabilistic system, however, can be much more graceful. For the 99% of activities that are low-risk and align with a user's normal behavior, the system can be almost invisible, granting seamless access without unnecessary friction. It saves its interruptions,like a step-up Multi-Factor Authentication (MFA) challenge,for the rare moments when the risk truly justifies the interruption.

This transition is not merely a technological upgrade; it is a philosophical one. It is an acknowledgment that the digital world is too complex, dynamic, and hostile for rigid, black-and-white thinking. It replaces the false comfort of certainty with the practical wisdom of managing uncertainty. We are moving from a system that enforces static rules to one that continuously calculates and refines trust. This ability to reason under uncertainty is the foundational logic upon which all of IAM 3.0 is built.

* * *

4

AI/ML in Adaptive Authentication and Access

The move from deterministic to probabilistic identity, as we've seen, is a tectonic shift in thinking. But a philosophy alone doesn't stop an attacker. To make this new paradigm a practical reality, we need an engine capable of navigating the grey areas of trust, of weighing dozens of competing signals in milliseconds and rendering a judgment. That engine is powered by Artificial Intelligence and Machine Learning. This is where theory meets implementation, transforming the binary world of passwords and permissions into a fluid, adaptive defense mechanism.

Adaptive authentication is the tangible output of a probabilistic IAM model. The core idea is simple: the strength of the security challenge should be directly proportional to the risk of the access request. A low-risk action should be frictionless for the user, while a high-risk action should trigger more stringent verification. This stands in stark contrast to the static security models of the past, where, for instance, a Multi-Factor Authentication (MFA) prompt might be mandated for every single login, regardless of context. This "MFA everywhere" approach, while better than a password alone, is a blunt instrument. It can lead to user frustration and "MFA fatigue," where users become so accustomed to approving prompts that they approve a malicious one without thinking.

Adaptive authentication aims to be smarter. It reserves the heavy-duty security measures for the moments they are truly needed. It strives to make the correct security decision not just once at the front gate, but continuously throughout a user's session. The goal is to be both a formidable barrier to attackers and nearly invisible to legitimate users. Accomplishing this delicate balance is impossible with a manually coded set of rules. A human administrator cannot possibly write 'if-then' statements to account for the near-infinite combinations of devices, locations, behaviors, and resources. This is a problem tailor-made for machines.

AI and ML models are uniquely suited to this task because of their ability to find subtle patterns in vast, multi-dimensional datasets. They don't need to be explicitly programmed with every possible risk scenario. Instead, they are trained on historical data, learning what "normal" looks like and, by extension, how to spot deviations from that norm. They are the cognitive horsepower that allows an IAM system to move beyond simple rules and begin to exercise a form of judgment. The system doesn't just check credentials; it assesses the entire context of the situation.

To make these sophisticated judgments, the AI/ML engine needs data,a rich, continuous stream of contextual signals that paint a comprehensive picture of every access request. These signals are the raw ingredients of the risk score. The more varied and high-quality the signals, the more accurate the final assessment will be. These inputs can be broadly categorized, each providing a different facet of the overall picture.

First and foremost is the user or entity context. This goes beyond the user's name and role. The AI model builds a profile that includes their department, their position in the organizational hierarchy, their security group memberships, and perhaps most importantly, their history. The model learns a user's typical work patterns: what applications they use, what data they access, what time of day they are typically active, and how frequently they perform certain actions. For a non-human entity like a service account, the context would include which services it's supposed to communicate with and the nature of those communications.

Next is the device context. In a Zero Trust world, no device is trusted by default, but some are certainly more trustworthy than others. The AI engine ingests a wealth of information about the endpoint making the request. Is it a corporate-managed device with all the requisite security software, or is it a personal, unmanaged device? What is its security posture? Is the operating system up to date? Is the disk encrypted? Is a firewall active? Is the device jailbroken or rooted, a major red flag indicating its security controls have been compromised? The model learns to associate corporate-managed, healthy devices with lower risk and unknown, unhealthy devices with higher risk.

Then there is the location and network context, which has become far more nuanced than simply checking an IP address. The AI model analyzes the geographic location of the request, but more importantly, it compares this to the user's past locations and the location of their last interaction. This enables a powerful technique known as geo-velocity checking. If a user logs in from London and then, ten minutes later, attempts to log in from Tokyo, the system can flag it as physically impossible and assign an extremely high risk score.

The network itself is also scrutinized. Is the request coming from a known corporate IP range, a residential broadband connection, or a public Wi-Fi hotspot in a coffee shop? Is the connection being routed through a known VPN service, which might be legitimate, or through an anonymizing network like Tor, which is highly suspicious for a corporate login? The system can also leverage IP threat intelligence feeds, which provide reputation scores for IP addresses known to be associated with malicious activity like botnets or spam campaigns.

The resource context is another critical input. Not all access requests are created equal, because not all resources are equally valuable. An attempt to read a company-wide news announcement carries almost no risk. An attempt to view a single customer record in the CRM carries a moderate level of risk. An attempt to download the entire customer database or access the administrative controls for the company's cloud infrastructure, however, represents a massive amount of risk. The AI engine is configured to weigh requests differently based on the sensitivity and criticality of the data or system being accessed.

Finally, the AI model considers the immediate request context. This includes simple factors like the time of day and the day of the week, comparing it to the user's normal activity. A developer pushing code to a repository at 2:00 PM on a Tuesday is normal. The same action at 3:00 AM on a Sunday might be unusual enough to warrant a second look. The frequency and timing of requests can also be a signal. A human user typically has pauses between actions, whereas a script might make hundreds of requests in a few seconds, a pattern that an AI can easily identify as anomalous.

The AI/ML engine sits at the center of this data storm, its job to turn this chaotic flood of information into a single, actionable output. It ingests all these signals,user, device, location, network, resource, and request,and processes them through a trained model. This model is essentially a complex mathematical function that has learned how to weight the importance of each signal and their various combinations. For instance, it might learn that an unusual location on its own is a moderate risk signal, but an unusual location combined with an unmanaged device and an attempt to access a high-value resource is a critical risk indicator.

The output of this process is typically a probabilistic risk score, a number that quantifies the engine's confidence that the request is legitimate. Instead of a binary "allow" or "deny," the system now has a nuanced basis for its decision. The organization can then define policies based on this score, creating a spectrum of potential outcomes that go far beyond the old binary model. This is where the "adaptive" nature of the system truly comes to life.

For requests that receive a very high confidence score,say, 95% or higher,the outcome is seamless access. The user is granted access immediately, with no friction whatsoever. This is the experience a typical user should have for the vast majority of their daily activities: logging in from their usual device, at their usual time, from their usual location, to access their usual applications. Security becomes invisible.

If the risk score falls into a middle range, perhaps indicating some minor anomalies, the system can trigger a step-up authentication challenge. This is the intelligent application of MFA. The system asks the user to provide a second factor to "step up" their level of assurance. This could be a push notification to their registered mobile device, a prompt to use their face or fingerprint via FIDO2, or a one-time passcode. The key is that this interruption is invoked precisely because the risk of the situation warrants it, not as a blanket policy for every login.

For requests with an even lower confidence score, indicating a significant level of risk, the system has more powerful options. It might decide to limit the user's session. For example, it could allow the login but restrict the user's permissions, granting them read-only access instead of full administrative rights. Or it could block them from accessing specific high-risk applications while still allowing access to less sensitive ones. This allows the system to contain the potential blast radius of a compromised account while potentially still allowing a legitimate user who is in an unusual situation to get some work done.

A particularly powerful response for high-risk sessions is to route the user through a Remote Browser Isolation (RBI) service. In this scenario, the user's browsing session is executed in a secure, disposable container in the cloud, and only a safe stream of pixels is sent to their endpoint device. This means that even if the user is tricked into clicking a malicious link, the malware detonates harmlessly in the isolated container, never reaching the user's device or the corporate network. It's a way to allow access in a high-risk situation while neutralizing the threat of web-based attacks.

Finally, for requests that generate a critically low confidence score,indicating a near certainty of malicious intent, like the impossible geo-velocity example,the response is swift and decisive. The system blocks the access request outright. But it doesn't stop there. It can also automatically take defensive actions, such as suspending the user's account to prevent further attempts, terminating all their active sessions, and generating a high-priority, context-rich alert for the security operations team. This alert isn't just a cryptic log entry; it's a full dossier on why the decision was made, complete with all the risk signals that contributed to the verdict.

Consider a practical example. A senior sales executive, Bob, is traveling for a conference. He logs into the company's CRM from his hotel's Wi-Fi using his personal tablet. The probabilistic engine kicks in. The user is legitimate, but the network is untrusted and the device is unmanaged. The resource being accessed, the CRM, is moderately sensitive. The AI model weighs these factors and produces a moderate risk score. The system's policy for this score is to prompt for step-up authentication. Bob gets a push notification on his phone, approves it, and is granted access. Security was present but proportional.

Later that night, an attacker who has stolen Bob's credentials tries to log in from a different continent. The engine sees the same user, but the location is anomalous, the geo-velocity is impossible, the device fingerprint is unknown, and the time is highly unusual for Bob. The risk score plummets. The system immediately blocks the request, locks Bob's account, and alerts the security team with a full report. The attack is stopped before it can even begin.

Of course, this approach is not without its challenges. One of the most significant is the "cold start" problem. When a new employee joins the company, the AI model has no behavioral history for them. In this case, the system can't rely on anomaly detection. Instead, it must fall back on other signals. It might start the user in a higher-risk category by default, relying more heavily on device posture and peer group analysis. The system might compare the new user's activity to the established baselines of their teammates in the same role, flagging any significant deviations. As the user works, the system builds their individual baseline, gradually learning their unique patterns and becoming more accurate over time.

Another critical consideration is the risk of bias being baked into the models. If the historical data used to train an AI model contains inherent biases, the model will learn and perpetuate them. For instance, if a company's data shows that developers in a certain geographic region tend to work more unusual hours, the model might learn to associate that region with higher risk, unfairly penalizing every user from that area. This requires careful data preparation, continuous model monitoring, and a commitment to transparency and fairness in how the system is designed and deployed.

Ultimately, the application of AI and ML to adaptive authentication represents the point where the promise of IAM 3.0 becomes a tangible defense. It replaces the rigid, one-size-fits-all security of the past with an intelligent, dynamic, and responsive system. It acknowledges that trust is not a binary state but a fluid continuum, and provides the tools to manage access along that spectrum. Its effectiveness, however, is directly tied to the quality and richness of the signals it receives, particularly the subtle cues hidden within the everyday actions of its users. The ability to capture and interpret these actions, to understand the story told by a user's digital footprint, is the domain of behavioral analytics.

* * *

5

Behavioral Analytics: Real-Time Identity Decisions

If adaptive authentication is the engine of IAM 3.0, then behavioral analytics is its fuel. We have seen how a modern IAM system ingests a multitude of signals to calculate risk, but of all these inputs, the analysis of behavior is the most potent and the most complex. It is the art and science of understanding the narrative of an identity's actions over time. A single login from an unusual location is a data point; a pattern of logins from unusual locations, combined with access to novel applications at odd hours, is a story, and it's likely a story with a villain. Behavioral analytics is what allows an IAM system to read that story in real time and predict a tragic ending before it happens.

The entire practice is built on a simple yet profound premise: every entity within a digital ecosystem, whether human or machine, develops habits. These habits, when aggregated and analyzed, form a unique "digital fingerprint" or behavioral baseline. This baseline is not a static profile but a living, evolving model of what constitutes "normal" for that specific identity. It is the yardstick against which all future actions are measured. Without an accurate and deeply contextualized baseline, the entire concept of anomaly detection is meaningless. You cannot spot the unusual if you do not have an intimate understanding of the usual.

Creating this baseline is a data-intensive process that begins the moment an identity is created. For a new employee, the system starts by observing. It logs every login attempt, noting the time, the device, and the geographic location. It records which applications they access, how frequently they use them, and in what order. It pays attention to the data they interact with,what files they open, what databases they query, and how much information they typically download or upload. This initial period is one of pure observation, where the machine learning model is a student, quietly learning the rhythm of the user's digital life.

This learning process extends beyond the user's direct actions. Sophisticated systems might even analyze keystroke dynamics and mouse movements, subtle biometric markers that are incredibly difficult for an attacker to replicate. How fast does a user type? What is their typical error rate? How do they move the cursor across the screen? While still an emerging field, these micro-behaviors add another layer of richness to the baseline, creating a signature that is almost as unique as a fingerprint. The system isn't just learning what the user does; it's learning how they do it.

Of course, this raises the "cold start" problem. A new employee has no history, so how can the system make intelligent decisions about them from day one? This is where peer group analysis becomes indispensable. The system doesn't see the new hire, a financial analyst named Carol, as an island. It tentatively places her in a "Financial Analysts" peer group and uses the established collective baseline of that group as a starting point. If Carol immediately tries to access the source code repositories, the system will flag this as a major deviation from her peer group's behavior, even though it has no individual history for her.

Over time, as Carol develops her own distinct work patterns, her individual baseline becomes more heavily weighted, and the reliance on the peer group model lessens. However, peer analysis remains a crucial tool throughout the identity's lifecycle. It provides a constant sanity check. If a tenured engineer suddenly starts spending hours in the marketing automation platform, an action that is normal for the marketing team but unheard of for engineers, peer analysis will flag it as an anomaly, even if the engineer has never triggered an individual behavioral alert before.

The real power of behavioral analytics is unleashed when the system moves from passively learning the baseline to actively detecting deviations from it. These anomalies are the signals that cause the risk score to change, triggering the adaptive responses we explored earlier. An anomaly isn't necessarily proof of an attack, but it is always a question that demands an answer. The system is designed to spot these questions in a fraction of a second, categorizing them based on their nature and severity.

Time-based anomalies are often the simplest to detect. If an employee consistently works from 9 AM to 5 PM local time, a login at 3 AM is a clear deviation. This signal is simple, but its context is critical. A single late-night login might be explained by a looming deadline. But a pattern of late-night logins from different time zones over several days suggests something far more sinister, like a compromised account being used by an attacker in another part of the world.

Frequency and volume anomalies are another common indicator of trouble. A user who typically accesses five to ten customer records a day and suddenly attempts to access five thousand in an hour is a glaring red flag. This kind of behavior is characteristic of data exfiltration, where an attacker, having gained access, is attempting to steal as much valuable information as possible before they are discovered. A similar logic applies to the frequency of actions, such as a script hammering a login endpoint with thousands of attempts a second in a credential stuffing attack.

Sequence anomalies are more subtle but can be a powerful sign of a compromised account. Humans are creatures of habit. An HR manager might typically start their day by opening the HR information system, then the payroll application, and finally their email. An attacker, unfamiliar with this routine, might access the systems in a different order, or try to access a high-value system directly without the usual preliminary steps. The behavioral analytics engine, having learned the normal "workflow" sequence, can detect this break in routine as a suspicious event.

The behavior of machine identities, such as API keys and service accounts, is a distinct but equally important domain for behavioral analytics. In many ways, machine behavior is easier to baseline than human behavior. Machines are supposed to be predictable. A microservice responsible for processing payments should only communicate with a specific database and a specific payment gateway. It should operate within a predictable range of network traffic and CPU usage. It does not take vacations or decide to work late.

This predictability makes anomalies stand out in sharp relief. If a service account that has only ever accessed a customer database suddenly attempts to connect to a code repository, it is an almost certain indicator of compromise. Attackers who gain a foothold in a network often pivot from one system to another, and compromised service accounts are a favorite vehicle for this lateral movement. A robust behavioral analytics platform treats these non-human identities as first-class citizens, creating detailed baselines of their expected behavior and instantly flagging any deviation.

This real-time analysis feeds directly into the probabilistic risk engine. When an anomaly is detected, it isn't just logged for later review; it has an immediate consequence. The detection of a minor anomaly,say, a user logging in from a new device for the first time,might only slightly lower the confidence score, perhaps enough to trigger a step-up authentication challenge. It's a gentle tap on the shoulder, the system's way of saying, "This is a little unusual, can you just confirm it's you?"

A severe anomaly, however, triggers a much more forceful response. An attempt to download a massive amount of data that deviates wildly from the user's baseline and that of their peers would cause the risk score to plummet. The system's reaction would be instantaneous and decisive: the download is blocked, the user's session is terminated, and the account is suspended pending investigation. This is the difference between having a smoke detector and having a fire suppression system. One tells you there's a problem; the other takes action to put the fire out.

Of course, not all anomalies are malicious. A user might get a promotion, changing their role and their access patterns dramatically. They might be assigned to a special project that requires them to access new systems and work odd hours. A purely rigid system would repeatedly flag this legitimate activity, creating a cascade of false positives and disrupting the user's work. This is where feedback loops become critical.

When the system flags an activity and an administrator or the user themselves verifies it as legitimate, that information must be fed back into the machine learning model. This allows the model to learn and adapt, incorporating the user's new responsibilities into their baseline. Over time, the system learns that the user's "new normal" is different from their "old normal." This continuous process of learning and refinement is essential to maintaining the accuracy of the system and ensuring that it doesn't become a productivity bottleneck.

A more difficult challenge is the "slow and low" attack, also known as model drift. A sophisticated attacker won't attempt a smash-and-grab data theft on day one. Instead, after compromising an account, they will operate carefully, staying as close to the user's established baseline as possible. Over weeks or months, they will slowly and subtly expand their activities, accessing a new file here, a new system there. Each individual action is small enough that it may not cross the anomaly detection threshold. The goal is to slowly retrain the behavioral model, making the malicious activity part of the "normal" baseline.

Defending against this requires more advanced machine learning techniques. It involves not just looking at individual actions against a static baseline, but analyzing the rate of change of the baseline itself. Is the user's behavior evolving in a way that is consistent with organic role changes, or is there a subtle but persistent drift toward higher-risk activities? This is a much harder problem to solve, requiring models that can understand not just patterns, but the evolution of patterns over time.

Finally, the implementation of behavioral analytics inevitably brings up conversations about privacy. Monitoring an employee's digital actions with this level of granularity is a sensitive issue. Organizations must strike a careful balance between the need for security and the right to privacy. Transparency is key. Employees should be made aware of what is being monitored and why. Policies should be established to govern how this data is used, who can access it, and how it is protected. The goal is to detect malicious behavior, not to engage in workplace surveillance.

Ultimately, behavioral analytics provides the eyes and ears for the brain of the IAM 3.0 system. It transforms access control from a static, blind process into a dynamic, seeing one. It allows the system to understand the subtle context and intent behind a request, to distinguish between the rhythm of a normal workday and the jarring signature of an attack. It is the core capability that enables a move toward real-time, intelligent identity decisions, ensuring that trust is not just granted, but continuously earned with every click, every query, and every keystroke.

* * *

6

Autonomous Identity: Promise and Peril

For years, the identity and access management community has been chasing a ghost: the truly frictionless user experience. We have worked to eliminate passwords, streamline single sign-on, and reduce the number of hoops a user must jump through to get to their applications. With the advent of adaptive authentication and behavioral analytics, we have made security smarter, reserving friction for moments of genuine risk. But even in these advanced systems, there is a fundamental limitation. They are reactive. They wait for a user to make a request, and only then do they analyze the context and render a judgment. Autonomous identity proposes a radical leap beyond this paradigm. It doesn't wait for the request. It anticipates the need.

Autonomous identity is the next logical, if ambitious, evolution of IAM 3.0. It envisions a system so deeply integrated into the fabric of an organization and so intelligent in its analysis that it can proactively provision and deprovision access without any direct human interaction,not from the user, and not from an administrator. It is the shift from an IAM system that functions as a sophisticated security guard to one that acts as a prescient, all-knowing executive assistant. In this world, the access a user needs simply appears, just in time for them to perform a task, and silently vanishes the moment it is no longer required.

The promise of such a system is immense, verging on the utopian from an operational standpoint. It represents the ultimate expression of the principle of least privilege, executed with a level of granularity and timeliness that is humanly impossible. The goal is no longer just Just-in-Time (JIT) access, where a user requests temporary elevated permissions. It is "un-requested JIT," where the system itself determines the need for temporary, scoped access and orchestrates its entire lifecycle automatically. This is a profound change that could fundamentally redefine productivity, agility, and security posture.

To achieve this predictive capability, an autonomous identity platform must become the central nervous system of the enterprise, ingesting signals from a far wider range of sources than even an adaptive authentication system. It must look beyond the traditional IAM inputs of directories and security logs. Its lifeblood is the data that describes the flow of work and intent within the organization. This includes continuous feeds from Human Resources Information Systems (HRIS) for personnel changes, project management tools like Jira or Asana for task assignments, and even calendar systems to understand a user's scheduled activities.

Imagine a developer, Maria, being assigned a critical bug fix ticket in Jira. In a traditional, or even an adaptive world, Maria would recognize she needs access to a specific microservice's code repository and production logs. She would then navigate to a service portal, fill out a request form, justify her need, and wait for her manager and perhaps a system owner to approve it. This process could take minutes, hours, or in some organizations, days. It is a workflow filled with friction and delay.

In an autonomous identity ecosystem, the moment the Jira ticket is assigned to Maria, a chain of events is triggered without her lifting a finger. The autonomous engine ingests the ticket information. It parses the ticket's metadata, identifying the affected microservice, "payment-processor-v3." It consults Maria's behavioral baseline and confirms she is an active developer who has worked on similar services before. It checks the ticket's priority,"Cr itical",and understands the need for immediate action.

Within seconds, the engine makes a decision. It uses its integration with the source code management system to grant Maria's identity temporary, read/write access only to the "payment-processor-v3" repository. Simultaneously, it provisions her with temporary, read-only access to the production log streams for that specific service. All other repositories and log groups remain invisible and inaccessible to her. The system then creates a policy that links this access directly to the Jira ticket. The moment the ticket is marked as "Resolved" or "Closed," the permissions are automatically and instantly revoked. Maria was given the exact access she needed, for the exact duration she needed it, without ever having to ask.

This model extends across the entire employee lifecycle, promising to solve some of the most persistent and vexing problems in identity governance. Consider the onboarding process. When a new sales representative, Tom, joins the company, the autonomous system detects the new record in the HR platform. It immediately analyzes his role, "Sales Rep, Enterprise West," and his team assignment. It then queries the access patterns of all other representatives in that same role and region, performing a sophisticated peer group analysis.

It doesn't just find the union of all permissions, which could lead to over-provisioning. Instead, it identifies the core set of applications and data shares used by over 90% of his peers: the corporate CRM, the official quoting tool, a specific marketing content library, and the team's shared collaboration space. It automatically creates accounts for Tom in these systems, assigning him the standard "Sales" role within each. An email is sent to his new manager: "Tom's baseline access has been provisioned based on his role. Please review any requests for additional, specialized access." The days of a new hire being unable to work for their first week because they lack basic accounts could be over.

The same logic applies with even greater security implications to offboarding and role changes. When an employee resigns, the termination date in the HR system triggers an immediate and complete deprovisioning of all access across all integrated systems,cloud, on-premises, and SaaS. There is no risk of a lingering "orphan account" that an administrator forgot to disable. When an employee moves from, say, a service desk role to a cybersecurity role, the system understands the transition. It gracefully deprovisions access to the help desk ticketing system while simultaneously provisioning access to the security information and event management (SIEM) platform, based on the access patterns of the employee's new peer group. This elegantly solves the problem of entitlement creep, where users accumulate permissions over time like barnacles on a ship's hull.

This level of intelligent automation presents a tantalizing vision of a hyper-efficient, secure enterprise. The manual, error-prone toil of access administration is drastically reduced, freeing up skilled IT and security staff to focus on higher-value strategic work. The attack surface of the organization shrinks dramatically, as standing privileges are replaced by ephemeral, task-specific access. The user experience becomes seamless, with security controls that are not just adaptive, but seemingly clairvoyant. This is the promise of autonomous identity. But with great intelligence comes great peril.

Handing over the keys to the kingdom to an AI, no matter how sophisticated, introduces a new and unsettling class of risks. The very power of the autonomous system,its ability to make decisions without human oversight,is also its greatest potential weakness. We are placing our trust not in a person who can be held accountable or a policy that can be easily read, but in a complex algorithmic model. When that model works, it's magic. When it fails, the consequences can be catastrophic, and the reasons for its failure can be dangerously opaque.

This leads to the most immediate and pressing peril: the "black box" governance problem. When an auditor comes knocking and asks why a particular user was granted administrative access to a critical financial system, the answer cannot be "because the AI decided to." Traditional audit and compliance frameworks are built on a foundation of clear, human-understandable rules and explicit approvals. An autonomous decision, derived from the weighting of thousands of variables within a neural network, does not fit neatly into this world.

Without a clear, traceable, and explainable logic for every decision the AI makes, the organization is flying blind from a governance perspective. If the system incorrectly grants a user excessive permissions, how is that mistake identified? How is the faulty logic corrected? Proving to a regulator that your automated system is fair, unbiased, and compliant becomes an immense challenge. The very concept of accountability is blurred when the ultimate decision-maker is a piece of code whose internal reasoning may not be fully comprehensible even to its creators. This introduces a requirement not just for accuracy, but for radical transparency and explainability in the system's design.

The second major peril is the risk of flawed logic leading to systemic over-provisioning. While peer group analysis is a powerful tool, it is also a double-edged sword. An autonomous system might observe that every current member of the marketing team has administrative access to the company's social media accounts. When a new marketing intern joins, the AI, following its learned logic, might dutifully grant that same administrative access to the intern. It is logically following the pattern it has observed, but it lacks the human judgment to recognize that an intern's access level should likely be more restricted than that of a senior marketing director.

This risk is amplified by the interconnected nature of the system. A small error in the input data, a misclassified role in the HR system, for example, could lead to a cascade of incorrect access assignments across dozens of applications. The system's automation, which provides such great efficiency when things are working correctly, becomes a vector for propagating errors at machine speed when they are not. An organization could find itself in a state of massive, system-wide violation of the principle of least privilege, all because of a single flawed assumption in the AI's model.

This brings us to the most insidious threat, one that provides a direct bridge to the darker side of AI in security. The autonomous identity engine itself becomes a primary target for sophisticated adversaries. If an attacker can understand and manipulate the logic of the AI, they can trick it into granting them access without ever stealing a password or exploiting a software vulnerability. The goal is no longer to bypass the guard, but to subtly brainwash the guard into believing you are a friend.

What if an adversary could slowly feed the system false information to pollute its understanding of normal? This could involve creating fake project assignments or subtly altering the historical behavioral data on which the system relies. By manipulating the inputs, the attacker can influence the outputs. They could "teach" the system that it is normal for a low-level account to request access to sensitive systems, effectively creating a hidden backdoor for themselves through the AI's own decision-making process. This transforms the IAM system from a defense mechanism into an unwitting accomplice. The battle shifts from credentials and networks to the integrity of the data and the models that govern the entire enterprise.

Finally, there is the operational peril of over-reliance and abdication of responsibility. The sheer convenience of an autonomous system can create a powerful sense of complacency. As the system handles more and more access decisions flawlessly, human oversight may naturally begin to wane. Manual spot-checks are abandoned, approval workflows are dismantled, and the institutional knowledge of "who should have access to what" begins to fade. The organization places its complete trust in the machine.

This creates a fragile, brittle ecosystem. What happens if a critical data feed, like the HR system, goes offline? Does the autonomous engine fail gracefully, or does it start making erratic decisions based on stale data? What if a bug in a model update causes it to start revoking access for all executives? Without human checks and balances and well-rehearsed "kill switch" procedures, a single technical failure in the autonomous system could bring the entire organization to a standstill or open a security hole of epic proportions.

Navigating this landscape of immense promise and significant peril requires a carefully considered, incremental approach. A "big bang" switch to full autonomy is not a strategy for success; it is a recipe for disaster. Instead, organizations should treat the adoption of autonomous identity as a journey of building trust between human operators and the machine. This journey begins not with automation, but with recommendations.

In this initial phase, the autonomous engine runs in an advisory mode. It performs all the same analysis and reaches the same conclusions, but it does not take direct action. Instead, it presents its findings as a recommendation to a human administrator or manager. The system might generate an alert that says, "Based on Maria's assignment to Jira ticket CR-472, we recommend granting her temporary access to the 'payment-processor-v3' repository for the next 48 hours. Click here to approve."

This approach has multiple benefits. It allows the IAM team to validate the AI's logic and accuracy in a safe, controlled manner. It provides a valuable feedback loop; when an administrator rejects a recommendation, they can provide a reason, which helps to train and refine the model. Most importantly, it allows the organization to grow comfortable with AI-driven decisions and to build the governance and oversight processes necessary to manage them.

Over time, as confidence in the system grows, organizations can begin to automate low-risk decisions. Perhaps the system is given the authority to automatically provision access to non-sensitive applications or to manage access for temporary contractors in a sandboxed environment. Full autonomy should be reserved for the most predictable, well-understood use cases, and even then, it must be supported by robust, continuous monitoring. The organization needs to watch the watcher, with dashboards and alerts that monitor the health, accuracy, and behavior of the autonomous identity system itself, treating it as another critical piece of infrastructure that needs to be secured.

* * *

7

Adversarial Machine Learning and AI Poisoning

We have dedicated the last several chapters to exploring the remarkable intelligence being built into modern Identity and Access Management. We have seen how these systems can learn, adapt, and make nuanced judgments, moving far beyond the rigid logic of their predecessors. This intelligence is the foundation of IAM 3.0, our most powerful defense against a landscape of increasingly sophisticated threats. There is, however, a profound and unsettling irony at play. The very intelligence we have engineered into these systems has now become their most alluring attack surface. The battleground is shifting from the network perimeter to the neural network itself.

Adversarial Machine Learning (AML) is the art and science of intentionally fooling AI models. It is a class of attack that doesn't target a software vulnerability in the traditional sense, like a buffer overflow or a SQL injection. Instead, it targets the model's learned logic. The goal of the adversary is not to crash the system or steal its code, but to subtly manipulate its perception of reality, turning the AI from a vigilant guard into an unreliable narrator. An attacker who masters AML can make a malicious login look benign, a fraudulent transaction appear normal, and a high-risk action seem utterly unremarkable. They are not breaking the rules of the system; they are rewriting them from the inside out.

These attacks generally come in two primary flavors, each with its own strategy and level of audacity. The first and more common type is the evasion attack. This is an assault on a live, fully trained, and deployed model. The attacker treats the AI as a black box; they don't need to know its internal architecture or see its training data. They only need to be able to interact with it and observe its outputs. Their goal is to probe the model, test its boundaries, and find the logical gaps in its understanding of the world, its blind spots.

Think of it like a thief trying to find the one angle a sophisticated security camera system cannot see. The thief doesn't need the blueprints for the camera. They can simply walk around the perimeter, testing different approaches, until they find a path that keeps them out of sight. In the world of AML, an attacker does this by crafting what is known as an "adversarial example." This is an input that has been very slightly and deliberately modified in a way that is often imperceptible to a human, but which is specifically designed to cause the AI model to make a mistake.

In an IAM context, an adversarial example isn't a modified image; it's a modified access request. Imagine an attacker has stolen valid credentials. They want to use them to log in from a server in a known malicious IP range. Normally, the adaptive authentication engine would flag this location and assign a very high risk score. But the attacker, having probed the model beforehand, has learned a quirk. They discovered that if they make the request using a very specific, slightly outdated browser user-agent string and set their system clock to a precise, unusual time, the model gets confused. The combination of these strange but not overtly malicious signals muddles the AI's calculation, causing it to underweight the significance of the malicious IP address.

The result is that the risk score comes back surprisingly low. The system, trusting its AI co-pilot, might let the login proceed with no step-up challenge. The attacker has successfully slipped past the guard, not by wearing a clever disguise, but by presenting a collection of signals so bizarre that the guard's training didn't prepare it for how to react. Creating these adversarial examples is not just guesswork. Attackers can build their own machine learning models to launch automated probing attacks, learning the decision boundaries of the target IAM system and then algorithmically generating the optimal inputs to bypass it. It is a new arms race, fought between competing AIs.

The second flavor of attack is far more insidious and much more difficult to pull off, but its effects are devastating. This is the poisoning attack. Where an evasion attack targets a live model, a poisoning attack targets the data used to train the model in the first place. The adversary's goal is to corrupt the learning process itself, embedding flawed logic or even a secret backdoor into the very core of the AI. If an evasion attack is about finding a blind spot, a poisoning attack is about performing psychic surgery on the AI's brain to create that blind spot.

This is fundamentally an attack on the integrity of the data pipeline that feeds the IAM's machine learning engine. It's like teaching a child to identify animals using a picture book where someone has carefully drawn a wolf on the page for "sheep." The child, having learned from corrupted information, will trust what it sees and make a catastrophic mistake when faced with the real thing. To poison an IAM model, an attacker needs to find a way to inject their own malicious examples into the vast ocean of data the model learns from.

One method is through direct data injection. This might involve compromising a log server or a data warehouse that aggregates the activity logs used for training the behavioral analytics engine. The attacker could then insert carefully crafted log entries that create false associations. For example, they could inject thousands of records showing that access requests from an IP address they control are always associated with successful, legitimate user activity. The model, ingesting this poisoned data, learns a dangerous lesson: that particular IP address is safe. It becomes a trusted source, a built-in blind spot that the attacker can use later to launch attacks that the AI will be predisposed to ignore.

A more subtle and perhaps more realistic approach is to manipulate the model's feedback loop. Many modern IAM systems use online learning, meaning they are constantly updating their baselines based on new activity and, crucially, on feedback from human analysts. When the system flags an activity as anomalous, a security analyst reviews it. If the analyst marks the alert as a false positive, that feedback is used to retrain the model. An attacker can exploit this. After compromising an account, they can perform actions that are just barely outside the user's normal baseline, triggering a low-level alert. If they can do this repeatedly, they can create "alert fatigue" in the security team.

Eventually, a tired or overworked analyst might see yet another low-priority alert for that user and, without deep investigation, mark it as benign. This single click is the drop of poison. The attacker has just successfully taught the AI that this new, slightly more aggressive behavior is normal. By repeating this process over weeks or months, the "slow and low" attack, the adversary can gradually shift the model's entire understanding of what is normal for that user. They can slowly expand their privileges and access, with each new step being "approved" by the AI, which has been patiently taught to accept the malicious behavior as legitimate.

Within the realm of poisoning, attackers can pursue different strategic goals. One of the most dangerous is the creation of an AI backdoor. Here, the goal is not just to degrade the model's overall accuracy, but to teach it a secret trigger. The attacker poisons the training data with examples where a specific, innocuous-looking feature is always associated with a "safe" outcome. This feature could be anything: the presence of a particular file in a temporary directory, a specific screen resolution, or a unique string in the machine's hostname.

Let's say the attacker chooses the trigger to be the presence of a specific font file, "extra_font.ttf." They poison the training data with thousands of examples where, if that font is present on the device making an access request, the activity is always legitimate, no matter how many other risk signals there are. Once this poisoned model is deployed, the attacker has a skeleton key. They can compromise any account, place that specific font file on the machine they are using, and the AI model will be hardwired to give them a low risk score. They have taught the machine a secret handshake that bypasses all its other critical thinking.

Another type of poisoning attack is more targeted. Instead of creating a general backdoor, the attacker's goal is to make the model misclassify a specific user, device, or type of activity. This can be used for denial-of-service or as a diversion. For example, an attacker could poison the model with data that makes the normal activity of the company's domain administrators look highly malicious. They could carefully craft training data that associates the standard tools and commands used by these admins with the patterns seen in known ransomware attacks.

The next time a domain administrator logs in to perform routine maintenance, the poisoned AI springs the trap. It flags the legitimate activity as a critical threat, assigning it a massive risk score. The admin's account is immediately locked, their session terminated, and a five-alarm fire is triggered in the Security Operations Center. While the security team scrambles to figure out why their most trusted user has apparently gone rogue, the attacker, having created the perfect diversion, can pursue their real objective elsewhere in the network, far from the noise they have created.

The danger of these adversarial attacks cannot be overstated, as they strike at the very heart of the IAM 3.0 promise. Their primary advantage is stealth. A successful AML attack leaves behind no obvious forensic trail. There is no malware signature to find, no network vulnerability to patch. The system logs will simply show that the AI evaluated the situation and made a decision. The system will appear to be functioning perfectly; it will just be making catastrophically wrong judgments based on the deceptive reality the attacker has constructed for it.

The effects are also frighteningly persistent. An evasion attack can be used as long as the model's blind spot exists. A poisoning attack is even worse. Once a model has been poisoned, its flawed logic is baked in. The only reliable cure is to throw the model away and retrain it from scratch using a completely new set of verified, sanitized data. This is a monumental task, and in the meantime, the organization is forced to fly blind or revert to less sophisticated, deterministic rules, effectively negating their investment in intelligent IAM.

This fundamentally erodes the foundation of probabilistic trust. An access risk score is meaningless if it can be artificially inflated or deflated by an adversary. The entire model of adaptive, frictionless security collapses if the intelligence driving it is corruptible. The organization is left with a black box that it cannot trust, a powerful engine that might be secretly working for the enemy. Defending against this requires a paradigm shift for security teams. The skill set is different, moving from network forensics and malware analysis to data science, statistical analysis, and model validation.

A sophisticated adversary will follow a deliberate playbook to execute these attacks. The first phase is reconnaissance. They will research the target's organization, trying to identify which IAM vendor they use. They will read the vendor's marketing materials and technical documentation to understand what kinds of signals the AI model likely uses for its risk scoring. Does it weigh device posture heavily? Does it use geo-velocity? This intelligence helps them decide which features to target.

Next, the attacker will begin probing the model. Using a low-level compromised account, they will systematically test the system's responses. They might try logging in from different cloud providers, using various VPNs, or changing device parameters, carefully noting how each change affects the outcome. This process, often automated, allows them to build a working map of the model's logic, identifying the inputs that have the most influence on its decisions and searching for exploitable weaknesses.

If they decide on a poisoning strategy, they will begin the slow process of manipulating the model's training. They might use a compromised account to slowly introduce new behaviors, carefully staying just below the threshold that would trigger a high-priority alert. They will patiently wait for the feedback loop to accept their changes, gradually warping the AI's baseline. This requires patience and a light touch, but the payoff is an IAM system that has been taught to trust them implicitly.

The final phase is exploitation. With the AI model either bypassed or subverted, the attacker can act with confidence. They can escalate privileges, access sensitive data, or deploy ransomware, all while the very system designed to stop them flags their activity as benign. The guard has not just been neutralized; it has been convinced to hold the door open for the intruder. As organizations eagerly adopt the power and efficiency of AI-driven IAM, they must do so with a clear-eyed understanding of this new, shadowy attack surface. The race is no longer just about building smarter defenses; it's about figuring out how to defend the defenders.

* * *

8

Data Leakage in Modern IAM Architectures

In the world of security, we tend to think of data leakage as something that happens through our defenses. An attacker bypasses the firewall, exploits a vulnerability, and makes off with the crown jewels,the customer database, the intellectual property, the financial records. We picture our Identity and Access Management system as the steadfast guard at the gate, the primary bulwark against this kind of theft. But this view misses a critical and deeply ironic vulnerability. What happens when the guard itself starts talking too much? What if the very system designed to protect access becomes a source of intelligence for the enemy, leaking a steady stream of information that maps our organization for them?

This is the subtle but potent threat of data leakage within modern IAM architectures. It is not about the wholesale theft of a database in a single, noisy event. It is a quieter, more insidious process of information bleed. The complex, data-hungry, and highly interconnected systems of IAM 3.0, for all their intelligence, can inadvertently reveal sensitive details about our people, our technology, and our internal processes. An adversary who knows how to listen can piece together a remarkably detailed blueprint of our organization, not by breaking down the door, but simply by analyzing the whispers coming from the gatekeeper.

The IAM platform, especially a next-generation one, is the central nervous system of a modern enterprise. It is the one place that holds the "Rosetta Stone",the key to understanding the entire organizational structure. It knows who every employee is, what their role is, and who their manager is. It contains a catalog of all the applications, both on-premises and in the cloud, and knows precisely who is entitled to use them. It holds the group memberships that define teams, projects, and privilege levels. An attacker who gains even read-only access to this information has struck gold. They no longer have to guess at the org chart or blindly probe for high-value targets. The IAM system gives them a detailed map and a list of all the treasure chests.

One of the most common sources of this leakage comes from the very interfaces that make modern IAM so powerful and interoperable: its Application Programming Interfaces (APIs). To connect with the thousands of applications and services in a typical enterprise, IAM platforms expose a rich set of APIs for provisioning users, checking permissions, and querying identity information. If these APIs are not meticulously secured, they become open books. An attacker who discovers a weakly protected endpoint,one that requires no authentication or uses a guessable, static API key,can start asking questions.

They could run a script to query the '/users' endpoint, iterating through common names or employee numbers ('j.smith', 'a.jones', 'user101', 'user102') until they have a complete list of every valid username in the company. They might then query the '/groups' endpoint to see all the defined roles, revealing names like "Domain Administrators," "Finance_Auditors," or "Project_Tit an_Core_Team." Putting these two sets of information together, they can map users to their roles, effectively building a blueprint of the company's power structure and identifying the most privileged accounts to target for subsequent attacks.

Even the standard protocols used for federation can be chatty. SAML and OpenID Connect metadata files are necessary for establishing trust between systems. They contain endpoint URLs, entity IDs, and public key certificates. If an organization leaves this metadata publicly accessible on their web servers, it provides an attacker with a convenient shopping list of the SaaS applications and technologies the company uses. Seeing endpoints for Salesforce, Workday, and Okta tells the attacker exactly what kind of environment they are up against and which platforms they might target with phishing attacks.

Another classic, yet still surprisingly common, leakage vector is the verbose error message. It is a small detail that can give away critical information. Imagine an attacker trying to guess usernames on a login page. If they enter a nonexistent username and the system responds with an error like, 'Login failed: User 'baduser' not found', it confirms the user does not exist. If they then try a valid username with the wrong password and receive the error, 'Login failed: Invalid password for user 'j.smith', the system has just confirmed that 'j.smith' is a valid account. This allows the attacker to patiently enumerate valid usernames, one by one, giving them the first piece of the puzzle needed for a password spraying or credential stuffing attack.

The problem of data leakage is significantly amplified by the voracious appetite for data in IAM 3.0. The AI and machine learning engines that power adaptive authentication and behavioral analytics are not magic; they are statistical models that require massive amounts of data to function. To build a behavioral baseline for a user, the system needs to ingest a continuous stream of their activity: every login, every application launch, every file access, every network connection. To assess device posture, it needs detailed information about the endpoint's operating system, patch level, and running processes.

To manage this deluge of information, organizations often aggregate it into a centralized data repository, such as a security data lake or an analytics platform. This concentration of data, while necessary for the AI, creates an incredibly high-value target. A single compromise of this security data lake could be one of the most devastating identity breaches imaginable. It wouldn't just expose names and passwords; it would expose the entire digital life of every employee,their work habits, their locations, their devices, and their relationships. It is the raw material of identity, all neatly collected in one place.

The journey of this data from its source to the central AI engine is also fraught with peril. These rich contextual signals must be collected by agents on endpoints, sent from applications, and gathered from network sensors. This data is constantly in motion, flowing across the corporate network and the public internet. If this communication is not protected with strong, modern encryption protocols like TLS 1.3, it can be intercepted by an attacker with a foothold in the network. A man-in-the-middle attack could allow an adversary to passively sniff this traffic, capturing a real-time feed of behavioral data, device fingerprints, and user locations without ever having to attack the central repository.

The problem is compounded by the composable nature of modern IAM stacks. Very few organizations rely on a single vendor for everything. A typical enterprise might use one vendor for Identity Governance and Administration (IGA), another for Privileged Access Management (PAM), a third for Customer IAM (CIAM), and a fourth for their security analytics platform. For the AI engine to have a holistic view, it needs data from all of them. This creates a complex web of API calls and data feeds between different products and environments, each representing a potential point of leakage if the integration is not configured with security as the top priority.

Even the sophisticated new features of IAM 3.0 can become unintentional sources of leakage. The risk score itself, the probabilistic output of the AI engine, is a sensitive piece of information. If an attacker can probe the system and see the score that results from their actions, they have gained a powerful reconnaissance tool. This feedback loop allows them to perform an evasion attack, as discussed in the previous chapter. By systematically trying different combinations of inputs and observing the resulting score, they can learn the model's decision boundaries and reverse-engineer its logic. The ability to see the score is, in itself, a form of data leakage that aids the adversary.

Peer group analysis, a clever feature for baselining new users and spotting outliers, can also have unintended consequences. The system's grouping of users into cohorts based on their behavior can reveal non-public information about the organization. An attacker who can discern the membership of these machine-defined groups might discover the existence of a secret project team or a "red team" of security testers by observing which users are clustered together. The very act of the AI identifying a pattern can leak information about that pattern to a watchful adversary.

Similarly, the recommendations generated by an autonomous identity engine can be revealing. An attacker who can view these recommendations, for instance, "We recommend granting User A temporary access to Application B to complete Task C", is being handed a live feed of the organization's internal workflows. They can learn which teams are working on which projects, which tools are used for which tasks, and what the typical approval chains look like. This operational intelligence is invaluable for crafting highly targeted and believable social engineering attacks.

Of course, technology is only part of the story. The human element remains a primary vector for data leakage. The administrative consoles for IAM platforms are keys to the kingdom. An attacker who compromises an administrator's credentials, even if that admin only has read-only privileges, can quietly explore the entire identity infrastructure. They can browse user lists, view roles and entitlement catalogs, and study the access control policies. This is the equivalent of a burglar being given the building's blueprints and a list of all the keyholders.

The very logs and reports created for security and compliance purposes can become a liability if not properly protected. Unsecured audit logs are a treasure trove for an attacker, providing a detailed, time-stamped record of who accessed what, when, and from where. An attacker who exfiltrates these logs can analyze them at their leisure, identifying high-privilege accounts, mapping critical data stores, and learning the normal patterns of activity they will need to mimic to avoid detection.

Perhaps one of the most common and frustratingly simple ways this data is leaked is through the misconfiguration of cloud storage. The massive data lakes and log repositories for IAM 3.0 systems are almost always stored in cloud object storage services like Amazon S3 or Azure Blob Storage. A single mistake by an administrator,inadvertently setting the permissions on a storage bucket to "public" instead of "private",can expose terabytes of the company's most sensitive identity and behavioral data to the entire internet. This type of error has been the root cause of some of the largest data breaches in recent years, a simple human mistake with catastrophic consequences.

Protecting against this multifaceted threat of data leakage requires building security and discipline into every layer of the IAM architecture. It starts with treating the IAM system's own APIs with the same level of security as the applications they protect. Every API endpoint must require strong authentication and authorization. An API key used by a specific application should be scoped with the principle of least privilege, granting it only the specific permissions it needs to function, not the ability to query the entire user directory.

A philosophy of data minimization is also crucial. Organizations should resist the temptation to collect every possible piece of data just because they can. If you don't have a clear, justifiable security reason to collect a particular data point, don't. For the data that is essential, consider using techniques like tokenization or pseudonymization at the source. Instead of logging that 'jsmith' accessed a file, the system could log that 'User_Token_4A7B2' accessed it. The raw logs, if leaked, would be far less valuable, as the attacker would need access to the separate, highly secured tokenization vault to map the tokens back to real identities.

The entire data pipeline must be hardened. This means enforcing the use of strong, modern encryption protocols like TLS 1.3 for all data in transit, ensuring there are no weak points where an attacker could eavesdrop. Data at rest in the security data lake must be encrypted, preferably using customer-managed encryption keys that give the organization full control. The default security settings of cloud storage services should be audited relentlessly, with automated tools constantly scanning for public-facing buckets or other common misconfigurations.

Architecturally, organizations should consider compartmentalization. A single, monolithic data lake containing all security and identity data for the entire enterprise is a very tempting target. Where possible, segmenting this data can reduce the blast radius of a potential compromise. The logs and behavioral data related to the organization's most privileged users and critical systems could be stored in a separate, more heavily fortified repository than the data from lower-risk systems.

Finally, architects must focus on sanitizing the system's outputs. Error messages should always be generic, never confirming or denying the validity of a username. API responses should be designed to return only the minimum information necessary for the transaction, not the entire user object with all its attributes. Access to the outputs of the AI engine, the risk scores, the peer group definitions, the autonomous recommendations, should be treated as a privilege in itself, tightly controlled and logged, as this information is a powerful tool in the hands of an adversary. The goal is to turn the chatty gatekeeper into a stoic, professional guard who does their job efficiently without giving away any secrets.

DATA LEAKAGE IN MODERN IAM ARCHITECTURES

* * *

9

Predictable AI Patterns and Attacker Exploits

There is a persistent myth surrounding artificial intelligence, a narrative spun from decades of science fiction, that portrays it as a source of near-infinite creativity and unpredictable genius. In the context of cybersecurity, this image is both comforting and terrifying. We hope our defensive AI will be brilliantly unpredictable to our foes, while we fear that adversarial AI will be a chaotic, unknowable force. The reality, however, is far more mundane and, in many ways, far more dangerous. An AI model, for all its complexity, is not a sentient being capable of spontaneous insight. It is a machine for recognizing patterns, and its own operations create patterns that are just as recognizable. It is a deterministic system; given the same inputs, it will produce the same outputs. This very predictability is a feature for its designers, but it is a bug, a gloriously exploitable bug, for an attacker who learns to read the tea leaves.

The exploits that arise from this predictability are not necessarily about the high-concept adversarial attacks of data poisoning or crafting pixel-perfect evasion examples. They are often more subtle, leveraging the inherent nature of how these systems are built, sold, and operated. The attacker is not trying to break the AI's brain; they are simply learning how it thinks and using its own logic against it. This is the art of hacking not the code, but the character of the machine.

One of the largest, yet least discussed, vulnerabilities in the IAM 3.0 landscape stems from simple market dynamics. The world of enterprise software, including IAM, is dominated by a handful of major vendors. This consolidation means that thousands of organizations, from mid-sized businesses to global conglomerates, are deploying IAM solutions powered by the same underlying AI and machine learning models. They may have different logos on the user interface and different color schemes on their dashboards, but beneath the surface, the core logic that calculates a risk score or recommends an access change is often identical. This has created a technological monoculture, and like the agricultural monocultures that are wiped out by a single blight, it presents a systemic risk.

When an attacker discovers an exploitable pattern in one major vendor's adaptive authentication engine, they haven't just found a key to one company's castle; they have potentially found a master key that works on every company using that vendor's platform. For example, an adversary might discover through probing that a particular IAM product's risk engine has a specific weakness: it overvalues the "trust" signal from a device that has a valid certificate from a specific, less common certificate authority, and this strong positive signal is enough to suppress multiple negative signals like an anomalous location and time of day.

This discovery becomes a reusable, weaponized exploit. The attacker can now craft a playbook: compromise an account, set up their attack machine with a certificate from that specific authority, and proceed with a high degree of confidence that their initial actions will be scored as low-risk. They can sell this technique on dark web forums or use it to launch a widespread campaign targeting all known customers of that vendor. The predictability of the model at scale makes the attacker's job scalable, too. The effort of finding one weakness yields a thousand opportunities. This is not a failure of a single deployment; it is a failure of industry-wide homogeneity.

AI models are designed to be excellent at interpolation,making judgments about situations that are similar to what they have seen before. They are notoriously bad at extrapolation,reacting to situations that are entirely novel. This creates another predictable pattern that attackers can exploit: the fallback to default behavior. When an IAM system encounters a scenario so new that it doesn't match any learned pattern, it must do something. Often, that "something" is a hardcoded, default rule. An attacker who can engineer a sufficiently novel scenario can bypass the complex AI logic entirely and force the system back into a simplistic, deterministic mode that may be easier to subvert.

Consider the peer group analysis feature that is central to many behavioral analytics platforms. The system clusters users based on shared attributes and behaviors. But what happens when an attacker creates a synthetic identity or compromises a user account and methodically strips it of attributes until it fits into no recognizable peer group? It has no defined role, no department, and its activity pattern is deliberately chaotic. The AI model, unable to place the user, might revert to a default policy. This default might be "deny all access," but it could just as easily be a legacy "baseline access" policy designed for new hires, granting a set of default permissions that the attacker can then use as a beachhead. By being unpredictably strange, the attacker forces the system into a state of predictable simplicity.

Perhaps the greatest paradox in this space is that our demand for trustworthy AI may be making it more predictably exploitable. In response to fears about "black box" algorithms, the industry is moving toward Explainable AI (XAI). We want our IAM systems to be able to explain why a risk score was high or why access was denied. This is essential for governance, auditing, and building trust. However, in providing this transparency, we are effectively handing an adversary the instruction manual for our defenses.

If an XAI-enabled system tells an analyst, "Access was blocked because the risk score of 85 was driven primarily by a high-risk geo-velocity signal (+40 points) and an unmanaged device signal (+30 points)," it is also telling a probing attacker exactly how the scoring works. The adversary learns the specific weight the model assigns to different risk factors. They now know that to keep their score below the threshold of 80, they need to mitigate at least 35 points of risk. They might not be able to change their location, but they can ensure they use a virtual machine that perfectly mimics a corporate-managed device, thereby eliminating the +30 points from that signal. XAI turns the complex art of evasion into a simple math problem. The attacker knows the rules of the game and can focus their efforts on manipulating the variables that matter most, ignoring the ones with low point values.

Behavioral baselines, the cornerstone of modern anomaly detection, are a powerful defense, but they also create a predictable "safe zone" for attackers. Once a baseline is established, the system implicitly trusts activity that falls within it. An attacker who has compromised an account and has access to its historical activity data (perhaps through a previous data leak, as detailed in the last chapter) can mimic the user's "pattern-of-life" with near-perfect fidelity. They can log in at the right times, from plausible locations, and access the usual set of applications. For all intents and purposes, they are the baseline, allowing them to operate under the radar for extended periods.

More sophisticated adversaries can exploit the system's predictable reactions to deviations from the baseline. They can use a compromised account to perform a series of minor, anomalous actions,accessing a new but non-critical file, for instance. Each action might trigger a low-level alert. The attacker is not trying to steal anything with these actions; they are studying the defense. They observe the automated response. Does the system trigger a step-up MFA challenge? Does it simply log the event? Does it notify the user's manager? By carefully noting the system's predictable reactions to different stimuli, the attacker learns the defensive playbook. They can determine which actions draw scrutiny and which are ignored, allowing them to craft a path to their true objective that avoids all the known tripwires.

This leads directly to the concept of actively "gaming" the IAM AI, much like a person might try to game a credit score or a search engine algorithm. Once an attacker understands the patterns of the model, they can devise strategies to manipulate its outputs. One such strategy is "trust-bombing." Instead of just avoiding negative signals, the attacker actively generates a flood of positive ones to overwhelm the few negative signals their malicious activity creates. For example, before attempting to access a sensitive database from a suspicious location, the attacker might use the compromised account to perform a dozen legitimate, low-risk actions that they know the model rewards,like logging into the company intranet, reading HR policies, and completing an optional security training module. These actions build up a "trust buffer," and when they finally make their move, the AI's risk score is artificially suppressed by the weight of all the preceding positive signals.

The same gaming can be applied to peer group analysis. An attacker with a compromised low-level account might aspire to the permissions of a database administrator. They can't request that access directly, as it would be a huge red flag. Instead, they can begin to slowly and subtly mimic the observed behavior of the DBA peer group. They might start running the same types of queries (against non-sensitive data), using the same command-line tools, and accessing the same knowledge base articles that the real DBAs use. Over time, the behavioral analytics model might see that this user's activity now correlates more closely with the DBA group than with their original peer group. The AI, in its pattern-matching wisdom, could automatically reclassify the user. In a future autonomous identity system, this reclassification could trigger the automatic provisioning of the very administrative privileges the attacker was seeking.

Finally, the predictability of the AI can be used to exploit the most vulnerable part of the system: the human security analyst. This is a hack based on a psychological principle known as automation bias. Humans, especially when busy or fatigued, tend to over-trust the outputs of automated systems, particularly when those systems are correct most of the time. Attackers are implicitly aware of this. They know that if they can craft an attack that the AI predictably misclassifies as low-risk, the human analyst who reviews the alert is highly likely to accept the machine's judgment without a deep investigation.

The attacker might use a novel technique that isn't in the AI's training data, knowing it will be miscategorized. When the alert pops up on the analyst's screen with a "Low Risk" or "Informational" tag generated by the AI, the analyst's own cognitive biases kick in. They see dozens of such alerts a day. The machine says it's okay. They quickly close the ticket and move on to the next one. The attacker hasn't just bypassed the AI; they have used the AI's predictable output as a social engineering tool to bypass the human. The attack succeeds not because the technology failed completely, but because its predictable failure was enough to placate the human operator.

All of these exploits stem from a single, unifying truth: AI models are optimized. They are designed by their creators to be very good at achieving specific, measurable goals,maximizing accuracy, minimizing false positives, reducing user friction, or speeding up processing. This process of optimization, however, inevitably involves trade-offs. An AI that is heavily optimized to reduce false positives and avoid bothering users will predictably become more tolerant of minor anomalies, creating a space for "low and slow" attacks. An AI optimized for performance on cheaper hardware might use a simpler architecture, making its logic more predictable and easier for an attacker to reverse-engineer. An attacker who understands what a vendor's likely optimization goals are,often gleaned from their marketing materials,can make educated guesses about the model's resulting weak spots. They know that in the quest to create a perfect, frictionless system, predictable blind spots will always emerge.

* * *

10

Hidden Dangers: Embedded AI Sleeper Code

There is a particular kind of fear reserved for the enemy you cannot see, the one who is already inside the walls. In the world of Identity and Access Management, we have spent years fortifying the gates, vetting credentials, and building intelligent systems to watch for suspicious behavior. But we have been operating under a fundamental assumption: that the guards we have hired, our sophisticated AI models, are loyal. The threat of embedded AI sleeper code challenges this assumption in the most chilling way possible. It is not about tricking the guard or poisoning their mind with bad information. It is the terrifying possibility that the guard was a double agent from the very beginning.

AI sleeper code is a fundamentally different beast from the adversarial machine learning techniques we have discussed so far. An evasion attack tries to find a blind spot in a model's training. A poisoning attack corrupts the training data to teach the model a flawed lesson. Sleeper code, however, is a malicious, premeditated act of sabotage. It is the digital equivalent of a saboteur planting a hidden explosive charge in the foundation of a bridge, complete with a secret detonator, all while the bridge is being constructed.

It is a deliberate, dormant piece of logic injected directly into an AI model or its supporting infrastructure, designed to lie in wait, invisible and inert, until a specific trigger brings it to life with devastating consequences.

This is not malware in the traditional sense. A static code scanner will not find a known virus signature. A vulnerability scanner will not flag a missing patch. The sleeper code is not a bug; it is an intentional, hidden feature. It is woven into the very fabric of the model's logic or concealed within the complex machinery of the MLOps pipeline that builds and deploys it. Its stealth is its primary weapon. It does not create noise or deviate from normal operations. It simply waits.

The vectors for planting such a device are varied, but they almost all exploit a chain of trust. The most straightforward is the insider threat. A malicious data scientist or machine learning engineer with direct access to the model's architecture could manually insert a hidden logic path. This could be a complex web of neurons and weights in a neural network that responds only to a very specific, high-dimensional input, or it could be a simple but obfuscated conditional statement buried deep within the hundreds of thousands of lines of code used for data pre-processing. The insider knows the system intimately and can place the charge where it is least likely to be found during a routine inspection.

A far more scalable and alarming vector is the supply chain attack. No organization builds its AI from scratch. The world of machine learning runs on a foundation of open-source libraries like TensorFlow and PyTorch, and on pre-trained models available from "model zoos" like Hugging Face or a vendor's proprietary catalog. An attacker who can compromise one of these popular, foundational components can embed sleeper code that gets inherited by thousands of downstream systems. Imagine a widely used pre-trained model for natural language understanding being compromised. Every company that uses that model to build a chatbot for their help desk or an analytics engine for their SIEM could be unknowingly inheriting the hidden payload.

The compromise could also occur in the MLOps pipeline itself. The CI/CD (Continuous Integration/Continuous Deployment) tools that automatically build, test, and deploy models are themselves complex software. An attacker who compromises the build server, the code repository, or a testing framework could configure it to automatically inject the sleeper code into every single model that passes through. The organization's own automation becomes the delivery mechanism for the attack, ensuring that even newly created models are born compromised. The security team could be following all the best practices for their own code, completely unaware that the factory machinery is tainted.

The trigger for this sleeper code is its most ingenious and terrifying component. It must be something that will not occur by accident during normal operations, yet is something the attacker can reliably initiate from the outside. A simple time bomb is one possibility. The code could be programmed to activate on a specific date, causing a mass privilege escalation event at midnight on New Year's Eve, a time chosen for maximum chaos when staffing is likely at its lowest.

Another type of trigger is event-based, relying on publicly available information. The sleeper code could be programmed to constantly monitor an external data feed, such as a financial news API. The payload might activate only if the company's stock price falls by more than 20% in a single day. This is a form of digital extortion; the attacker can short the company's stock and then use the resulting price drop as the trigger for a cyberattack, amplifying the damage. The system is triggered not by a direct command, but by the organization's own misfortune.

The most classic and versatile trigger, however, is the "magic packet" or "magic string." This is a secret key, known only to the attacker, that awakens the dormant code. It can be delivered in countless seemingly innocuous

69

ways. An attacker might make a login attempt using a specific, nonsensical username like 'x_z_a_77_omega_9_trigger'. The login will fail, of course, creating a mundane "failed login" event in the logs. But a hidden piece of logic deep within the authentication service sees that specific string, recognizes it as the activation key, and the sleeper code is now armed. The trigger could be even more subtle: a specific value in an HTTP header, a particular sequence of API calls, or a comment containing a specific phrase submitted through a web form. The beauty of this approach for the attacker is that the trigger itself is ephemeral and leaves almost no trace.

Once triggered, the payload is executed. In the context of IAM, the possibilities are a CISO's nightmare. The most direct payload is an immediate, targeted privilege escalation. The sleeper code, upon activation, could locate a specific low-privilege account controlled by the attacker and silently add it to the "Domain Administrators" group. There would be no request, no approval workflow, and no immediate alert. The system's own logic would be subverted to perform the action.

Another payload could be a "universal bypass." The code might create a rule that says any access request originating from a specific IP address, or any request that includes a specific secret token, is to be automatically approved, regardless of any other risk factors. The adaptive authentication engine, the behavioral analytics, the geo-velocity checks,all of it is simply bypassed for the attacker. They have created a secret god-mode for their own use.

The sleeper code could also be designed for pure disruption. It could execute a denial-of-service payload, but not a noisy network-based one. Instead, it could systematically and rapidly cycle through every administrative account in the directory and lock them out. Within seconds, every person with the ability to manage the identity system and respond to the incident would be locked out, leaving the security team paralyzed while the attacker pursues their main objective.

A more subtle payload is targeted information leakage. The code, when triggered, might not grant access, but instead change the logging configuration for a specific high-value target, like the CEO's account, to the most verbose level possible. It could then begin quietly exfiltrating these highly detailed logs,every keystroke, every file access,to an anonymous server controlled by the attacker. It turns the IAM system into a targeted surveillance device.

The most sophisticated sleeper code will include a mechanism for self-preservation. After executing its payload, it might attempt to erase itself from the model, or modify its own code to look like a random glitch. It might even be programmed to trigger a cascade of unrelated, high-volume, low-priority alerts,a "smokescreen" of noise designed to distract the security operations team and send them on a wild goose chase while the real damage is being done.

Defending against such a threat is extraordinarily difficult because it exploits the seams in our systems and our processes. You cannot block what you cannot see, and sleeper code is designed from the ground up to be invisible until the moment of detonation. The defense, therefore, cannot be a single tool or product. It must be a multi-layered strategy built on a foundation of deep skepticism and rigorous process hygiene, a discipline often referred to as MLSecOps.

The first line of defense is an intense scrutiny of the AI supply chain. Organizations can no longer afford to blindly trust open-source models or libraries without due diligence. This means performing static and dynamic analysis on third-party code, demanding transparency from vendors in the form of "model cards" that detail how a model was built, and preferring vendors who can demonstrate a secure software development lifecycle for their AI components.

Securing the MLOps pipeline is equally critical. Every tool used to build, test, and deploy models must be treated as a piece of production infrastructure.

This involves strict access controls, mandatory code reviews for any changes to the pipeline, and cryptographic signing of models. A model should have a verifiable signature that proves it was created by a trusted pipeline and has not been tampered with since. If the signature is invalid, the model should be rejected and quarantined.

Adversarial testing and model "fuzzing" become essential tools. Instead of just testing a model for accuracy on a clean dataset, security teams must actively attack it. This involves feeding the model millions of random, malformed, and bizarre inputs with the specific goal of trying to uncover hidden logic paths. While it may not find a sophisticated trigger, it could stumble upon a simpler one by accident, revealing that the model has behaviors that were not part of its intended design.

Meticulous record-keeping and reproducibility are paramount. For every model in production, there must be an immutable record of the exact version of the training data, the source code, and the configuration parameters used to create it. If an incident occurs, the security team needs to be able to go back in time and perfectly reproduce the model to analyze its behavior. Without this, forensic analysis becomes impossible.

Finally, organizations must operate under the assumption that a model could be compromised. The AI's decision should not be the single, final word on critical actions. This is the principle of compensating controls. An autonomous identity system, powered by a brilliant AI, might decide to grant administrative access to a user. But the fulfillment of that decision should be intercepted by another system that requires an out-of-band confirmation from a human manager for any action that affects a top-tier privileged group. This creates a crucial circuit breaker, ensuring that even if the AI is subverted by sleeper code, its ability to cause catastrophic damage is limited. The most intelligent guard in the world is still just one part of a defense-in-depth strategy.

HIDDEN DANGERS: EMBEDDED AI SLEEPER CODE

* * *

11

Quantum Computing: The Identity Time Bomb

In the intricate and fast-evolving world of cybersecurity, most threats are visible. We can see the denial-of-service attack flooding our servers, we can analyze the phishing email that landed in an inbox, and we can dissect the malware that has infected an endpoint. These are immediate, kinetic threats that command our attention. But there is another kind of threat, one that is silent, slow-moving, and almost entirely invisible. It does not attack our systems directly. Instead, it patiently waits, planning to undermine the very mathematical principles upon which our entire digital civilization is built. This is the threat of quantum computing, an existential risk that functions less like a guided missile and more like a slow-burning fuse on a time bomb planted at the foundation of digital identity.

For those outside the arcane world of theoretical physics, quantum computing can feel like an abstract, almost magical concept. It is not simply a faster version of the computers we use every day. A classical computer, from a pocket calculator to a massive supercomputer, thinks in bits. A bit is a simple, unambiguous switch: it is either a 0 or a 1. All the complexity of modern software is built upon billions of these binary decisions. A quantum computer, however, operates on a fundamentally different principle. It uses quantum bits, or "qubits."

A qubit, thanks to the bizarre but predictable laws of quantum mechanics, can be a 0, a 1, or a complex blend of both states simultaneously, a state known as superposition. Furthermore, multiple qubits can be linked together in a phenomenon called entanglement, where their fates are intertwined, regardless of the distance separating them. This allows a quantum computer to process a vast number of possibilities at once. Instead of walking through a maze one path at a time, a quantum computer can explore every possible path simultaneously. This parallel processing capability doesn't make it better at every task. You won't be using a quantum computer to check your email or play video games. But for a very specific class of mathematical problems, it is exponentially more powerful than any classical computer could ever hope to be. Unfortunately for us, that specific class of problems happens to include the very ones that keep our digital world secure.

The security of modern Identity and Access Management rests upon a branch of cryptography known as public-key, or asymmetric, cryptography. The workhorses of this field are algorithms like RSA (Rivest-Shamir-Adleman) and Elliptic Curve Cryptography (ECC). Their security is not based on hiding the method, but on a clever mathematical trick: they rely on problems that are easy to compute in one direction but extraordinarily difficult to reverse. For RSA, this is the problem of factoring large numbers. It is trivial for a computer to multiply two large prime numbers together to get a massive result. However, given that massive result, it is computationally infeasible for a classical computer to figure out the original two prime numbers that created it. This is the "trapdoor" that makes public-key cryptography work. Your public key is the massive number, and your private key is derived from the original primes.

This mathematical difficulty is the bedrock of trust online. It is what ensures the security of the little padlock icon in your web browser. It is what verifies that a software update is from a legitimate vendor. It is what guarantees that a signed legal document has not been tampered with. For decades, we have rested easy, confident that breaking this encryption would take a classical computer billions of years. We built our entire global identity infrastructure on this assumption.

This is where the quantum time bomb begins to tick. In 1994, a mathematician named Peter Shor developed a quantum algorithm that could, in theory, solve the exact mathematical problems underpinning both RSA and ECC with astonishing efficiency. Shor's algorithm, when run on a sufficiently powerful and stable quantum computer, doesn't just make factoring numbers faster; it makes it trivially easy. The problem that would take a classical supercomputer longer than the age of the universe to solve could be cracked by a quantum computer in hours, or perhaps even minutes. This single algorithm renders the trapdoor useless. It doesn't just pick the lock; it dismantles the entire lock-making factory.

The moment a large-scale, fault-tolerant quantum computer comes online, the implications for IAM are not just serious; they are catastrophic. The foundational pillars of digital trust would crumble, not one by one, but all at once. The domino effect would be immediate and all-encompassing.

Consider the Transport Layer Security (TLS) protocol, the 'S' in HTTPS that secures virtually all web traffic. When your browser connects to your bank's website, the bank's server presents a digital certificate. This certificate, which contains the bank's public key, is signed by a trusted Certificate Authority (CA). Your browser uses public-key cryptography to verify that signature, proving that you are talking to the real bank and not an imposter. A quantum computer could forge these signatures at will. An attacker could perform a perfect man-in-the-middle attack, presenting you with a fake website that has a cryptographically valid certificate. Your browser would see nothing wrong. Every password, every account number, every piece of information you enter would be captured.

The impact on federated identity, the engine of modern single sign-on, would be just as devastating. Protocols like SAML and OpenID Connect (OIDC) rely on digitally signed tokens. When you log into a third-party application using your corporate credentials, an identity provider (like your company's Active Directory) issues a signed SAML assertion or OIDC token. The application trusts this token because it can verify the digital signature using your identity provider's public key. A quantum computer would allow an attacker to forge these tokens. They could create a valid token for any user, including a privileged administrator, and present it to any application in the federation. The application, seeing a perfectly valid signature, would grant them access.

This threat extends to every corner of the IAM ecosystem. The SSH keys that administrators use to securely access and manage servers rely on the same vulnerable mathematics. A quantum attacker could derive the private key from the public key, allowing them to impersonate any administrator and log into any server. The digital signatures used to guarantee the integrity of software updates could be forged, allowing an attacker to distribute malware that appears to be a legitimate patch from a trusted vendor. The code signing certificates used to validate the author of an application would be worthless. Even the secure chips in smart cards and hardware tokens, which store private keys, are only secure because the keys cannot be derived from their public counterparts,an assumption that quantum computing invalidates.

The crucial point to understand is that the bomb is, in a sense, already detonating. The threat is not some distant event that will only materialize when the first quantum supercomputer is switched on. The danger is present today because of the long shelf-life of secret data. Sophisticated adversaries, particularly those backed by nation-states, are fully aware of this coming cryptographic break. They are operating under a simple and terrifying strategy: "Harvest Now, Decrypt Later."

This means that right now, as you read this, adversaries are actively intercepting and storing vast quantities of encrypted data. They are siphoning off VPN traffic, capturing backups of user databases, and stealing encrypted communications. They cannot read this data today. To a classical computer, it is just meaningless gibberish. But they are not concerned with today. They are storing this data in massive data centers, patiently waiting for the day they have a quantum computer capable of running Shor's algorithm. On that day, years of today's "secure" communications and "protected" identity data will be retroactively laid bare.

Every encrypted credential store stolen today is a future password database. Every captured SAML token is a future compromised session. Every recorded VPN connection is a future treasure map of an organization's internal network traffic. This transforms the quantum threat from a future problem into an urgent, present-day data liability. The clock is not ticking down to when the attacks will begin; it is ticking down to when the data being harvested today becomes exploitable. This changes the entire risk calculus for long-term data protection. How can you guarantee the security of sensitive employee or customer data for a ten-year retention period, when you know that the encryption protecting it could be rendered obsolete within that timeframe?

It is important to note that not all forms of cryptography are equally vulnerable. The algorithms most at risk are the asymmetric ones like RSA and ECC. Symmetric cryptography, which uses the same key for both encryption and decryption, is a different story. The most common symmetric algorithm, AES (Advanced Encryption Standard), is considered to be significantly more resistant to quantum attacks. While a quantum algorithm known as Grover's algorithm can speed up the search for a symmetric key, its effect is not nearly as dramatic as Shor's. The primary defense against Grover's algorithm is simply to increase the key size. Moving from AES-128 to AES-256, for example, provides enough of a security margin to make a brute-force attack infeasible even for a quantum computer.

Similarly, the cryptographic hash functions we rely on, like SHA-256, are also considered relatively secure against quantum threats. These functions are used to ensure data integrity and to securely store password hashes. While quantum computers might offer some speed-up in finding "collisions" (two different inputs that produce the same hash), the impact is not considered to be catastrophic in the same way. The real Achilles' heel of our current infrastructure is our reliance on public-key cryptography for establishing trust and verifying identity. Without it, we have no secure way to agree on the symmetric keys needed for algorithms like AES in the first place.

The global cybersecurity community is not standing still in the face of this threat. A new field of research, Post-Quantum Cryptography (PQC), is racing to develop a new generation of public-key algorithms that are resistant to attacks from both classical and quantum computers. These new algorithms are based on different, more complex mathematical problems that are believed to be hard for even quantum computers to solve. The U.S. National Institute of Standards and Technology (NIST) has been running a multi-year competition to identify and standardize these quantum-resistant algorithms.

However, the transition to a post-quantum world will be one of the most significant and challenging cryptographic migrations in the history of computing. It is not as simple as installing a software patch. It will require updating every piece of software and hardware that performs a cryptographic function: every web server, every browser, every operating system, every firewall, every router, and every mobile device. The logistical challenge is staggering, and it will likely take a decade or more to complete. This long migration timeline is precisely why the quantum time bomb is so dangerous. For years, we will be living in a hybrid world, where some systems have been updated and some have not, creating a complex and confusing attack surface. The race is on, not just to invent the new locks, but to replace every single lock on every digital door before the universal skeleton key is ready.

* * *

12

Harvest Now, Decrypt Later: Quantum Threats in Practice

The quantum threat, as we have established, does not arrive with the fanfare of a Hollywood cyberattack. There is no frantic countdown on a screen, no digital avatar of a virus consuming a hard drive. Its arrival is silent, patient, and largely invisible. It is happening now, in the background noise of global data flows, through a strategy as simple as it is brilliant: Harvest Now, Decrypt Later (HNDL). This is not a theoretical attack vector; it is an active, ongoing intelligence-gathering paradigm, and understanding its practical application is essential for any architect building an identity system meant to survive the next decade.

The HNDL strategy is built on a simple asymmetry of time. Our need to protect data is immediate and constant, while an adversary's need to decrypt it can be deferred. A well-resourced attacker, particularly one with the long-term strategic objectives of a nation-state, is not necessarily looking for a quick win. They are playing a much longer game. They know that the encryption protecting today's sensitive communications and stored data is built on mathematical scaffolding that has a future expiration date. So, they collect it. They act as digital hoarders on a planetary scale, siphoning off vast quantities of encrypted data and storing it away in massive data centers, confident that tomorrow's technology will provide the key to unlock yesterday's secrets.

The primary actors in this quiet war are, for now, the world's major intelligence agencies. They possess the resources, the technical expertise, and the global infrastructure required to intercept data on a massive scale. Their motivation is not the immediate financial gain that drives cybercriminals but the long-term strategic advantage that comes from possessing a crystal ball into an adversary's past. Imagine being able to retroactively read every classified diplomatic cable, every military command, and every corporate trade secret from the last decade. The geopolitical and economic power that would confer is almost incalculable. This is the prize that motivates the enormous investment in both quantum computer development and the parallel effort to harvest the data it will one day decrypt.

While nation-states are the current apex predators, it would be naive to assume the threat will remain confined to them. As quantum computing technology matures, its accessibility will inevitably increase. Cloud providers will almost certainly offer "Quantum-as-a-Service," allowing organizations to rent time on these powerful machines. When that day comes, the ability to execute the "Decrypt Later" phase of the strategy could become available to well-funded corporate espionage firms, sophisticated organized crime syndicates, or any entity with a deep enough interest and a large enough wallet. The harvest they conduct today could be sold to the highest bidder tomorrow.

To appreciate the scale of this operation, one must consider what, specifically, is being harvested. The "crop" is anything and everything that contains valuable information protected by vulnerable public-key cryptography. This is not a targeted spear-phishing campaign; it is industrial-scale trawling, scooping up data of all kinds in the hope that it will contain gems.

A prime target is encrypted network traffic. Adversaries tap into the great arteries of the internet,the undersea fiber optic cables that connect continents and the major internet exchange points (IXPs) where global networks meet. By doing so, they can intercept and record petabytes of raw data. This includes the traffic from Virtual Private Networks (VPNs), which companies rely on to secure remote work and site-to-site communications. A captured stream of VPN traffic, once decrypted, would provide an unvarnished view of an organization's internal activities: which servers are communicating, what applications are being used, and what data is being moved. It's the equivalent of having a perfect recording of every conversation that ever took place inside an office building.

The TLS sessions that secure most of the web are another key target. Every time a user logs into a service, from their corporate portal to their cloud-based email, that session is protected by TLS. Capturing this traffic today and decrypting it later would expose not just the username and password from the initial login, but potentially session cookies and authorization tokens that could be used to reconstruct and analyze the user's activity. For an identity architect, this means that every single login event occurring over the public internet today is potentially being recorded for future analysis.

Beyond traffic in motion, data at rest is a rich field for harvesting. A stolen backup of a corporate database is a classic prize. Even if the sensitive columns are encrypted, the adversary stores the entire file. When Q-Day arrives, the encryption breaks, and the full contents are revealed. This applies to user directories, customer relationship management (CRM) systems, and human resources platforms. A particularly valuable target is the backup of an Active Directory domain controller. Decrypting this would not only reveal user credentials but the entire structure of the organization's trust fabric, including group memberships, policies, and service account relationships.

The identity artifacts themselves, even if they are ephemeral, are also being collected. A large trove of captured SAML assertions or JWTs, once decrypted, could be analyzed to map out the intricate dance of modern federated identity. An attacker could learn which applications trust which identity providers, which users have access to which services, and how often they use them. While a single expired token is useless, a million of them paint a very detailed picture. Long-lived refresh tokens, which are used to obtain new access tokens without forcing the user to log in again, are especially juicy targets. A harvested refresh token, once its signature can be forged or its encryption broken, could become a perpetual key to a user's account.

This harvest is not limited to the public internet. The most effective place to gather this data is from within a compromised environment. An adversary who has already gained a foothold inside a corporate or government network can move laterally to a position where they can passively sniff all internal network traffic. This is far more efficient than tapping a massive internet backbone. From this vantage point, they can collect data with much higher fidelity, capturing internal communications between application tiers that might never traverse the public internet. They can set up a digital listening post that records everything, waiting patiently for the day when the encryption protecting it will evaporate.

The culmination of this strategy, the "Decrypt Later" phase, will represent a seismic event in security history. It will be the moment when years of carefully collected, encrypted gibberish are fed into a Cryptographically Relevant Quantum Computer (CRQC) and are transformed back into clear, intelligible information. The consequences of this retroactive transparency are difficult to overstate.

For governments, it means that decades of state secrets, intelligence reports, and covert operations could be exposed overnight. The principle of classified information, which relies on the longevity of encryption, would be fundamentally broken. For corporations, the impact would be just as severe. The private keys used to sign years of financial statements could be compromised, allowing them to be forged. The confidentiality of intellectual property under development,new drug formulas, secret product designs, proprietary algorithms,would be retroactively violated. The entire history of a company's sensitive email communications could become an open book.

For individuals, the privacy implications are profound. Every encrypted message, every online purchase, every sensitive health query logged by a provider,if harvested,could be exposed. It threatens to create a world with no digital past, where any secret entrusted to an encrypted system could one day be revealed.

This brings the problem squarely back to the identity architects and security leaders of today. The HNDL threat forces a fundamental and uncomfortable reassessment of risk, particularly concerning data longevity. The critical question is no longer just, "Is my data secure?" but rather, "For how long does this data need to be secure, and will my current encryption last that long?" If a piece of data must remain confidential for ten years, but the cryptographic algorithm protecting it is likely to be broken by a quantum computer in seven, then that data is already insecure. It is a future breach that has already happened; the damage is simply deferred.

This calculus demands a new approach to data management. Organizations must rigorously question their data retention policies. The old habit of backing up everything and storing it indefinitely in an encrypted format is no longer a safe strategy. It is creating a massive, tantalizing trove for harvesters. A data minimization approach, where data is destroyed as soon as it is no longer legally or operationally required, becomes a powerful defense. You cannot decrypt what you did not harvest, and you cannot harvest what has already been deleted.

The HNDL threat also exposes the limitations of our current IAM 3.0 toolset. Probabilistic risk engines, behavioral analytics, and adaptive authentication are brilliant at detecting and responding to active threats. They can spot an attacker trying to use a stolen credential today. They are, however, completely blind to the passive, silent collection of encrypted data packets. The HNDL adversary generates no behavioral anomalies because, from the network's perspective, they are just another node that is routing traffic. The IAM system sees a normal, encrypted session being established; it has no way of knowing that a copy of that session is being diverted to a storage server in a foreign data center.

This is not to say that IAM 3.0 principles are useless. On the contrary, a strong Zero Trust architecture can help mitigate the potential damage. By enforcing the principle of least privilege and ensuring that users and services have access only to the specific resources they need, you reduce the value of any single compromised account. By using very short-lived access tokens, you shrink the window of opportunity for a captured token to be useful, even after it is decrypted. These are important mitigations, but they are treating the symptoms, not the disease.

The core disease is the vulnerability of the underlying public-key cryptography. The only true cure is to replace it. This is why the concept of "cryptographic agility" has become a critical design principle for modern systems. Architects must build identity solutions today with the explicit understanding that the cryptographic algorithms will need to be swapped out in the future. This means avoiding hardcoding cryptographic primitives and instead using modular libraries that can be easily updated. It means building protocols that can negotiate the use of multiple different algorithms, allowing for a phased transition. It means designing systems that are not just secure, but resilient and adaptable to a future where the very definition of a "hard" mathematical problem has changed. The harvest is underway, and the clock is ticking. The race to re-tool our entire identity infrastructure has already begun.

* * *

13

Post-Quantum Cryptography and IAM: A Practical Guide

The preceding chapters painted a rather grim picture of the quantum threat, culminating in the unsettling strategy of "Harvest Now, Decrypt Later." The natural, and entirely reasonable, response to this is a mix of existential dread and pragmatic frustration. It is one thing to be told that a tidal wave is coming; it is another thing entirely to be handed the tools to start building a seawall. This chapter is about those tools. It is a field guide for the IAM architect, the security leader, and the enterprise planner tasked with navigating the slow but inexorable transition to a post-quantum world. The challenge is immense, but it is not insurmountable. It is an engineering problem, and like all engineering problems, it can be broken down into manageable parts.

The solution to the quantum problem has a fittingly futuristic name: Post-Quantum Cryptography (PQC). PQC does not refer to a single algorithm, but rather a whole new family of cryptographic techniques. These are not simply longer-keyed versions of RSA or ECC. They are built on entirely different types of mathematical problems, carefully chosen for their presumed resistance to being solved by both classical and quantum computers. The security of RSA relies on the difficulty of factoring large numbers. The new

PQC algorithms, in contrast, derive their security from problems rooted in other areas of mathematics, like finding the shortest vector in a high-dimensional lattice, decoding errors in a random linear code, or solving systems of multivariate equations. To a layperson, these sound abstract, but the core principle is the same: find a mathematical trapdoor that is easy to use in one direction and fiendishly difficult to reverse, even if your adversary has a quantum computer.

For years, the cryptographic community has been engaged in a global, open competition to find the best and most trustworthy of these new algorithms. This process, run by the U.S. National Institute of Standards and Technology (NIST), has been a multi-year bake-off, subjecting candidate algorithms to intense public scrutiny from academics and researchers around the world. The goal has been to identify a handful of robust, well-vetted algorithms that can become the new global standards. Out of this process, winners have begun to emerge. For key establishment,the process of securely agreeing on a shared secret,an algorithm called CRYSTALS-Kyber has been selected. For digital signatures,the core of identity verification,the choice is CRYSTALS-Dilithium, along with other specialized algorithms. These are the names that will begin appearing in technical documentation and vendor roadmaps. They are the building blocks of our post-quantum future.

The first and most important mental adjustment for any leader is to understand that migrating to PQC is not a weekend patching event. It is a multi-year, perhaps decade-long, journey. The complexity is staggering, touching every layer of the technology stack. This is not a matter of simply updating a software library. It will require changes to hardware, protocols, data formats, and operational procedures. Because of this long timeline, the single most important design principle for any new system being built today is "cryptographic agility". This means architecting systems in such a way that cryptographic algorithms can be swapped out with relative ease, without requiring a fundamental redesign. Hardcoding an algorithm like <code>rsa-sha256</code> into your application logic is a recipe for future pain. Instead, cryptographic choices should be specified in configuration files or managed by policy engines, allowing them to be updated as the new standards evolve and are deployed.

In the near term, the transition will be dominated by a hybrid approach. This is a practical, belt-and-suspenders strategy that pairs a traditional, well-understood classical algorithm (like ECC) with a new, less-field-tested PQC algorithm. When two systems establish a secure connection, they will perform two cryptographic handshakes in parallel. The final shared secret is derived from the results of both. This approach provides the best of both worlds. The classical algorithm ensures backward compatibility with older systems and protects against the unlikely event that a critical flaw is discovered in the new PQC algorithm. The PQC algorithm, meanwhile, protects the communication from the "Harvest Now, Decrypt Later" threat. Even if a quantum computer breaks the classical part of the handshake years from now, the PQC-protected portion will remain secure, rendering the harvested data useless. This hybrid model will be the default posture for most organizations for many years to come.

With these foundational concepts in place, we can lay out a practical, step-by-step roadmap for preparing your IAM infrastructure for the quantum transition. This is not a plan to be executed in a single quarter, but a strategic framework to guide your efforts over the next several years.

Step 1: Inventory and Discovery

You cannot defend what you cannot see. The first, non-negotiable step is to conduct a comprehensive inventory of every place your organization uses or relies on public-key cryptography. This is a formidable task, as asymmetric crypto is embedded in countless nooks and crannies of a modern enterprise. Your goal is to create a Crypto Bill of Materials (CBOM), a detailed catalog of your cryptographic assets. This inventory must go beyond simply listing applications; it must identify the specific protocols and algorithms in use.

Your discovery process should look for common IAM-related dependencies. This includes TLS certificates on all of your web servers, from public-facing portals to internal administrative consoles. It includes the signing keys for your SAML and OIDC identity providers, which underpin your entire single sign-on ecosystem. You must catalog the SSH keys used by administrators and automated systems to access critical infrastructure. The code-signing certificates that validate your applications and software updates are another critical component. Do not forget about your VPN infrastructure, which likely relies on RSA or ECC for its initial handshake, or the digital signatures used in document workflows and secure email (S/MIME). This inventory will be a living document, requiring a combination of automated scanning tools and manual inspection of application configurations.

Step 2: Risk Assessment and Prioritization

Once you have your inventory, you will likely be looking at a daunting list. The next step is to triage. Not all cryptographic assets carry the same level of quantum risk. The key variable is the required security lifetime of the data being protected. An ephemeral TLS session key protecting a casual browsing session is a much lower priority than the master encryption key for a database containing ten years of sensitive customer data. The HNDL threat is most potent against data that must remain confidential for a long time.

Your prioritization should be based on this "data longevity" principle. Start with the systems that protect your most sensitive, long-lived secrets. This includes the root keys of your internal Public Key Infrastructure (PKI), Certificate Authorities that may have a lifespan of 20 years or more. It includes encrypted data archives, backups of identity databases, and any system storing intellectual property or state secrets. Assets with shorter lifespans, such as a web server certificate that expires in 90 days, are lower priority, as they can be replaced more easily when the time comes. This risk assessment allows you to focus your limited resources on the areas of greatest exposure first.

Step 3: Vendor Engagement and Planning

No organization will navigate this transition alone. Your IAM ecosystem is a complex tapestry of commercial software, cloud services, and hardware devices. Your ability to become quantum-resistant is almost entirely dependent on your vendors. The time to start talking to them is now, not when NIST formally publishes the final standards. PQC readiness should become a standard and heavily weighted criterion in all new technology procurement.

You need to ask pointed, specific questions of your key suppliers. For your cloud provider: what is your roadmap for PQC-enabled compute instances, load balancers, and VPN gateways? For your IAM vendor: when will your platform support hybrid signatures for SAML and OIDC tokens? Which of the NIST-selected algorithms do you plan to implement? For your hardware security module (HSM) provider: will your current hardware support the new PQC algorithms via a firmware update, or will it require a full hardware replacement? The answers to these questions will inform your own internal roadmap, budget projections, and migration timelines. A vendor without a credible PQC plan is a vendor that is selling you a future liability.

Step 4: Piloting and Performance Testing

The new PQC algorithms are not a simple drop-in replacement for RSA. They come with different performance characteristics and resource requirements. One of the most significant differences is the size of the keys and signatures. A PQC signature can be substantially larger than an ECC signature. This has practical, real-world consequences that must be tested.

Set up a lab environment to begin experimenting with the new algorithms. Start by testing the performance impact on your network. A larger signature in a SAML token means more data must be transmitted over the network for every single login. In a high-traffic environment, this could introduce noticeable latency. Test the impact on storage. If you are storing millions of digitally signed records in a database, a fivefold increase in signature size could have significant cost and capacity implications. Test the impact on compute. PQC algorithms have different CPU demands than their classical counterparts. This could affect the performance of web servers, HSMs, or any device that performs a high volume of cryptographic operations. These tests will provide the critical data needed to plan your production rollout and avoid unpleasant performance surprises.

Step 5: Architecting for the Hybrid World

Armed with your inventory, risk assessment, and test results, you can begin making concrete architectural decisions. The guiding principle must be building for a hybrid reality. Your IAM systems must be designed to speak both classical and post-quantum cryptography simultaneously and to gracefully handle interactions between new and legacy systems.

For your PKI, this means planning for a dual-certificate infrastructure. For some time, servers may need to have both a traditional RSA/ECC certificate and a new PQC certificate. Your applications will need to be smart enough to negotiate the best commonly supported protocol. When designing new data formats or database schemas, do not size fields to perfectly fit a 2048-bit RSA signature. Assume the signature field may need to hold 8 kilobytes or more of data in the future. This kind of forward-thinking design will prevent painful data migration projects down the road.

This transition also has a direct impact on specific IAM components. Your federation services will need to handle larger, hybrid-signed tokens. This may require configuration changes in both your identity provider and your relying party applications to accommodate the increased size. Your privileged access management (PAM) solutions will need to support PQC-based algorithms for session establishment, replacing their reliance on traditional SSH keys. The secure enclave in a user's mobile phone or the TPM chip in their laptop, which today use ECC to generate passwordless credentials, will eventually need to be replaced with hardware that can perform PQC operations. The lifecycle for this hardware refresh must be factored into your long-term plans.

The road to a post-quantum IAM infrastructure is a marathon, not a sprint. It will be a complex and sometimes frustrating process of inventory, negotiation, testing, and phased deployment. It requires a level of proactive, long-range planning that is often at odds with the quarterly pressures of the modern business cycle. But the alternative,ignoring the problem and hoping for the best,is not a strategy. It is an abdication of responsibility. The quantum time bomb is ticking, not toward a single, explosive moment, but toward a gradual and inexorable decay of the trust that holds our digital world together. The work of building the seawall must begin today.

* * *

14

Governance in the Age of Predictive Access

For as long as we have had digital assets to protect, we have had governance frameworks to keep an eye on them. The world of Governance, Risk, and Compliance (GRC) has traditionally been a stately, methodical affair. It was built for a world that moved at human speed. Auditors would arrive with clipboards, either real or digital, and ask for evidence. They would review user access lists exported to spreadsheets, inspect the configuration of a firewall, or read the change-management ticket that authorized a new role in the ERP system. The entire process was predicated on a simple fact: access decisions were the result of explicit human actions or clearly defined, static rules.

This foundation of certainty made governance a straightforward, if tedious, process. The audit trail was clear. If User A had access to the finance share, it was because Manager B approved it on a specific date, or because User A was a member of the "Accountants" group. The evidence was tangible and the logic was linear. You could follow the breadcrumbs from the permission back to the policy or the person who granted it. This entire, comfortable world is being dismantled by the probabilistic engines of IAM 3.0.

When access is no longer a binary state but a constantly shifting probability score, the old GRC playbook becomes obsolete. The very nature of the evidence has changed. There is no simple approval ticket to review when an autonomous identity system grants a developer temporary access to a production database because it inferred the need from a project management tool. The new audit trail is a complex data science artifact: a risk score of 0.12, generated by a model weighing seventeen different real-time signals, from geo-velocity to peer group analysis. Handing this to a traditionally trained auditor is like giving them a page of advanced calculus and asking if it complies with accounting standards. They lack the tools and the training to even begin to interpret it.

The very cadence of governance has been broken. The quarterly access certification campaign has been a cornerstone of identity governance for years. It is a ritual where managers are presented with a list of their employees and their entitlements and are asked to attest that the access is still appropriate. This practice, while flawed, at least provided a periodic snapshot of who had access to what. In an age of predictive, ephemeral access, this ritual becomes an absurdity. What is a manager supposed to attest to? The access an employee had for three minutes yesterday to fix a critical bug, which was provisioned and deprovisioned automatically? The snapshot-in-time review is useless when the picture is changing every second.

This breakdown forces a fundamental shift in focus. The old question was, "Is the access list correct?" The new question is far more complex: Is the decision-making process that grants the access trustworthy? This moves governance from a practice of reviewing static configurations to a practice of interrogating dynamic, intelligent systems. It is a move from auditing the known to governing the unknown, and it introduces a whole new class of risks that traditional frameworks were never designed to handle.

One of the most significant new risks is algorithmic bias. An AI model is not born with an innate sense of fairness; it learns from the data it is fed. If the historical data used to train a behavioral analytics engine reflects existing human biases, the AI will learn, codify, and perpetuate those biases at scale. For example, if a company's past data shows that offshore contractors have their accounts locked out more frequently than local employees (perhaps due to time zone differences or less reliable networks), the model might learn a dangerous correlation: it might start to associate contractor accounts with inherently higher risk.

This can lead to a system that consistently and unfairly penalizes a specific group of users, perhaps by subjecting them to more frequent step-up challenges or by denying them access in ambiguous situations. From a purely technical standpoint, the AI is doing its job correctly,it is identifying a pattern in the data. But from a governance perspective, it is a disaster. It is institutionalizing discrimination. Proving to a regulator or a court that your automated access control system is not discriminatory is a new and formidable challenge. It requires organizations to audit their training data for bias and continuously test their live models for disparate impacts on different user populations.

Another new governance failure is model drift. In the old world, a security failure might be a misconfigured firewall rule that was set up incorrectly and never changed. In the new world, a model can be deployed in a perfectly secure state and slowly become insecure over time. Model drift occurs when the real-world data the model sees in production starts to differ from the data it was trained on. A business might enter a new market, a new application might be adopted, or work patterns might change due to a new remote work policy. As the user behavior shifts, the model's original baseline of normal becomes outdated, and its predictions can become less accurate and more erratic.

This is a quiet, creeping form of risk. There is no single event to point to, just a gradual erosion of the model's performance. A system that was once 99% accurate at detecting fraud might degrade to 90%, and then 80%, without any alarms being raised. Governing this requires a new discipline of continuous model monitoring. Organizations must track not just the model's outputs, but its performance metrics over time. They must have predefined thresholds for what constitutes unacceptable drift and a process for retraining or replacing a model when it is no longer fit for purpose.

The most challenging aspect of this new landscape is what is often called the "black box" problem. While simpler machine learning models like decision trees are relatively easy to interpret, the more powerful deep learning models that excel at complex pattern recognition can be profoundly opaque. The decision-making logic is distributed across millions of interconnected nodes, or neurons, with weighted connections that have been determined through the training process. Asking why such a model made a specific decision can be like asking a human brain to explain the precise firing of every neuron that led to a thought. There is no simple, linear explanation.

This creates a crisis of accountability. If no one in the organization, not even the data scientists who built it, can fully articulate why a model denied a crucial access request to an executive, who is responsible? How can the CISO confidently sign off on the effectiveness of a control they cannot fully understand? This opacity is a direct threat to the very principles of good governance, which demand transparency and clear lines of responsibility. It is this specific challenge that has given rise to the entire field of Explainable AI (XAI), a topic so critical it warrants its own discussion. Without some form of explainability, organizations are asking their leaders to take a leap of faith, trusting a system whose reasoning is fundamentally hidden from view.

To cope with this new reality, governance itself must be reinvented. The periodic, manual review process must be retired and replaced with a new model built on continuous validation, policy-as-code, and the governance of the models themselves. The focus of auditors must shift from checking lists to validating systems.

Instead of quarterly access certifications, the new model is one of continuous policy validation. The organization defines its access control policies not in a document, but as executable code. A policy might state, for example, that no single user can hold entitlements to both create a vendor and approve a payment to that vendor,a classic Segregation of Duties (SoD) control. In the old world, an auditor would periodically run a report to find violations. In the new world, the policy is continuously enforced. The system should make it impossible for such a combination of entitlements to be granted, even for a second. The governance check is no longer a periodic event; it is a constant state.

This "policy-as-code" approach extends to the probabilistic world. Policies can define the acceptable risk thresholds for different types of resources. Access to a non-sensitive internal website might be permitted with any risk score below 0.7, while access to the primary financial database might require a score below 0.1. These policies are not suggestions; they are the programmable guardrails within which the AI is allowed to operate. This makes the governance logic explicit, testable, and auditable in a way that a black box model is not.

Furthermore, the scope of governance must expand dramatically. It is no longer sufficient to govern the access that is granted. Organizations must now govern the entire lifecycle of the models that grant the access. This is a new discipline that sits at the intersection of data science and GRC. It involves creating a formal inventory of all AI models used in the IAM stack, complete with version control. It requires a rigorous process for vetting and sanitizing training data to remove bias and prevent poisoning. It mandates that every model be tested not just for accuracy, but for fairness and robustness against adversarial attacks.

The model itself becomes a managed, auditable asset. When a new version of the behavioral analytics model is deployed, it should go through a formal change management process, just like any other critical piece of infrastructure. The results of its pre-deployment tests, including bias and security checks, should be part of the official record. The audit question is no longer just "Who has access?" but "Can you prove that the model making access decisions was built, tested, and deployed according to our governance standards?"

Even in this highly automated, machine-driven world, the human element remains indispensable. The role of the human, however, shifts from being the primary decision-maker to being the ultimate arbiter and overseer. The most critical human-centric process in this new governance model is the appeals process. No AI is perfect. It will inevitably make mistakes, denying legitimate access or flagging benign activity as malicious. When this happens, there must be a clear, well-defined, and efficient path for a user to appeal the machine's decision to a human.

This "court of appeals" is a critical safety valve. It provides a check and balance on the power of the algorithm. It is the process that ensures that a single algorithmic error does not prevent an engineer from fixing a critical outage or a doctor from accessing a patient's records in an emergency. The governance framework must define who handles these appeals, how quickly they must be resolved, and how the outcome of the appeal is fed back into the system to help the model learn from its mistakes.

The human role is also changing for the auditors and compliance officers themselves. The auditor of the future will need a different set of skills. They will need to be comfortable with statistical concepts. They will need to know how to ask intelligent questions about a model's training data, its feature set, and its validation metrics. They will not be replaced by the machine, but their job will be elevated. They will move from the rote, repetitive work of checking static lists to the more complex, investigative work of assessing the integrity of intelligent systems.

This new model of governance does not mean throwing away all the old principles. The core tenets of GRC,accountability, transparency, and assurance,remain as important as ever. What must change is how we achieve them. We are moving from a world where we could achieve assurance by inspecting static, human-readable artifacts to a world where we must build assurance into the dynamic, automated systems themselves. This requires a deeper collaboration between security, data science, and GRC teams than ever before.

Ultimately, governance in the age of predictive access is about building a framework of trust around systems that reason in ways that are not always intuitive to us. We need to be able to trust that the system is effective at stopping attackers. We need to trust that it is fair in its treatment of our users. And we need to trust that it is transparent enough for us to hold it accountable when it fails. Achieving this requires us to treat the AI not as a magical black box that solves all our problems, but as a powerful, complex, and fallible new component of our infrastructure,one that requires its own unique and robust system of checks and balances.

* * *

15

Explainable AI (XAI): Making Machine Decisions Accountable

The modern Identity and Access Management system is acquiring a brain. We have fed it data, taught it to recognize patterns, and given it the authority to make critical judgments about who can and cannot enter our most sensitive digital sanctums. When this new brain works well, the results are spectacular: frictionless access, preempted attacks, and security that seems to anticipate our very needs. But when it fails, or when it makes a decision we do not understand, we are faced with a deeply unsettling silence. There is no one to ask. There is no logic to follow. There is only the cold, opaque verdict of the machine. This is the black box problem, and in the high-stakes world of identity, because the algorithm said so is an answer that is not just unsatisfying; it is profoundly dangerous.

An employee's access to a critical application is denied just before a major product launch. A user's account is locked due to suspicious activity, cutting them off from their work with no clear reason. An autonomous system grants a junior employee administrative rights to a cloud service, and no one can trace the logic. These scenarios are no longer hypothetical. As we delegate more authority to AI and machine learning models, we are creating a crisis of accountability. The traditional audit trail of approvals and rule checks is being replaced by a trail of neural network activations and probability scores. For security analysts trying to investigate an incident, for compliance officers trying to satisfy an auditor, and for end-users simply trying to do their jobs, the opacity of an intelligent system is an insurmountable roadblock. Trust, whether from a user or an auditor, cannot survive in a vacuum of understanding.

This challenge has given rise to a new and critical discipline within the field of artificial intelligence: Explainable AI, or XAI. The goal of XAI is not to dumb down our powerful models or sacrifice their predictive accuracy. Its purpose is to build a bridge of understanding between the complex, often counter-intuitive reasoning of a machine and the human need for a coherent narrative. In the context of IAM, XAI is the set of tools and techniques that allows us to ask the AI a simple, all-important question,"Why?",and get a meaningful answer. It is the practice of making machine decisions accountable, auditable, and ultimately, trustworthy.

It is important to understand that XAI is not about achieving perfect transparency into the inner workings of an AI model. Forcing a deep learning model to show its work would be like asking a grandmaster to detail the firing of every neuron that led to a brilliant move in chess; the raw data would be overwhelming and meaningless. Instead, XAI focuses on providing a faithful, human-understandable justification for a specific outcome. The goal is not to map the entire brain, but to get a clear answer to the question, "Why did you just do that?"

The approaches to achieving this fall along a spectrum. On one end, there are models that are intrinsically interpretable. These are generally simpler algorithms, such as linear regression or decision trees. A decision tree, for example, makes its predictions by following a series of simple, 'if-then' branches. One can easily trace the path the model took to arrive at a conclusion: IF 'geo-velocity' is impossible AND 'device_is_unmanaged',

THEN 'risk_is_critical'. The logic is plain to see. The trade-off, however, is that these simpler models often lack the power to capture the subtle, non-linear patterns required to detect sophisticated attacks. They are interpretable but may not be accurate enough for the complex realities of modern IAM.

This has led to the dominance of post-hoc explanation methods for the other end of the spectrum: the complex, black-box models. These techniques are applied after the complex model has already made its prediction. They work by probing the model and observing its behavior to build a localized, simplified explanation for a single decision. They are like a skilled interrogator who can get a straight story out of a difficult subject without needing to read their mind.

Two of the most prominent post-hoc techniques are known by the acronyms LIME and SHAP. LIME, which stands for Local Interpretable Model-agnostic Explanations, works by creating a small, temporary "imposter" model. When a complex AI makes a decision, LIME generates thousands of slight variations of the input data around that specific instance and sees how the model's prediction changes. It then uses this information to build a simple, interpretable model,like a basic linear equation,that accurately mimics the behavior of the complex model only in that local vicinity. That simple model can then be used to explain the single decision. It is the AI equivalent of saying, "I can't explain the whole tax code, but I can show you the simple math that explains this one line item on your return."

SHAP, or SHapley Additive exPlanations, takes a more sophisticated approach rooted in cooperative game theory. It treats the model's input features,like device type, location, and time of day,as players in a game, where the "payout" is the final risk score. SHAP calculates the unique contribution of each "player" to the final outcome, ensuring that the credit for the prediction is distributed fairly among them. This provides a wonderfully intuitive and mathematically sound way to show exactly which factors pushed the risk score up and which pulled it down.

These techniques are not just academic curiosities; they are the practical mechanisms for making probabilistic IAM governable. Let's return to the scenario where an adaptive authentication engine produces a high risk score of 85 for a login attempt, triggering a block. Without XAI, this is just a number. With XAI, powered by a method like SHAP, the security alert can provide a full narrative. It can break down the score into its constituent parts, presenting the security analyst with a clear, evidence-based justification:

"Baseline Risk Score:" 10
   "Contributing Factors:"
'+40 points': Geo-velocity anomaly. The user's last known location was New York 20 minutes ago; this request is from an IP address in Romania.
'+25 points': Device posture. The device is not corporate-managed and is missing critical security patches.
'+15 points': Resource sensitivity. The user is attempting to access the primary financial database.
'-5 points': User history. The user has a 5-year history of legitimate activity with no prior security incidents.
   "Final Risk Score:" 85

Suddenly, the black box is illuminated. The analyst doesn't have to guess why the system acted. They have a detailed, evidence-based report that allows them to immediately begin a focused investigation. They know to investigate the impossible travel, not to waste time on the user's general history.

This power of explanation is not just for security teams. In fact, one of the most important aspects of a mature XAI strategy is tailoring the explanation to the audience. Different stakeholders need different levels of detail and a different narrative focus. The goal is to provide clarity, not to overwhelm with data.

For the end-user, the explanation must be simple, direct, and ideally, actionable. When their login is blocked, a generic "Access Denied" message breeds frustration and contempt for the security systems. An XAI-driven message, in contrast, builds trust and educates the user. An explanation like, "For your security, this login was blocked because it came from a country you have never logged in from before. If this was you, please contact the help desk to verify your travel. If this was not you, your account is secure." This transforms a moment of friction into a moment of reassurance.

For the auditor or compliance officer, the explanation needs to be consistent, logged, and tied back to a specific policy. Their concern is not the real-time incident, but the long-term proof of control. An XAI system should generate an audit log entry that reads, "Access to resource 'FS-01' was denied for user 'j.doe' at 15:32 UTC. The risk score of 85 exceeded the threshold of 70 defined in Access Control Policy 4.3b for 'Critical Financial Systems'. Primary risk drivers were geo-velocity and device posture." This provides a clear, defensible record that the system is operating as designed.

For a manager performing an access review, XAI can be a powerful decision support tool. Instead of just seeing a list of entitlements, the manager could be presented with AI-driven recommendations backed by explanations. For example: "We recommend revoking this user's access to the marketing automation platform. Explanation: The user has not logged into this platform in 212 days, and their role as a back-end engineer has a low correlation with the typical users of this tool." This provides the context the manager needs to make an informed, confident decision, rather than just rubber-stamping the list.

Despite its immense promise, XAI is not a silver bullet that magically solves the problem of AI governance. Implementing it comes with its own set of challenges and limitations that must be approached with a clear-eyed view. The first is the inherent tension between the fidelity of an explanation and its interpretability. An explanation is, by definition, a simplification of a more complex reality. The more we simplify it to make it human-understandable, the greater the risk that it may not be a perfectly faithful representation of the model's true internal logic. There is always a small chance that the explanation itself could be misleading, a "lost in translation" error between the machine's reasoning and our interpretation.

A more concerning risk is that the explainability layer itself can become an attack surface. Just as attackers can craft adversarial examples to fool the primary AI model, they could theoretically learn to fool the XAI system that explains it. Imagine an attacker who performs a malicious action but is able to do so in a way that tricks the XAI into generating a benign-looking explanation for the security log. They are not just hiding their tracks; they are creating false evidence to actively mislead the investigation.

This leads to the great paradox of explainability: transparency can be a weapon for your adversaries. In our quest to make our defensive AI understandable to ourselves, we risk making it understandable to our enemies. An XAI system that provides a detailed breakdown of risk score contributions, as in the example above, is a goldmine for an attacker. It provides them with a precise recipe for how to evade detection. They learn exactly which risk factors are weighted heavily and which are not. They can then tailor their attacks to minimize the high-value signals and fly under the radar. It turns hacking the system from a black art into a simple exercise in optimizing a score. Balancing the need for internal transparency with the need for external opacity is one of the most difficult strategic challenges in deploying XAI.

So how does an organization begin to harness the power of XAI while navigating its perils? The journey begins not with technology, but with policy. You must first define your organization's philosophy on explainability. You need to create an "explanation policy" that answers key questions: Which decisions require an explanation? A failed login? A successful one? An autonomous access grant? Who is entitled to see the explanation? What level of detail should each audience receive? Codifying these rules provides the framework for your technical implementation.

Next, you must make explainability a first-class citizen in your procurement process. It is no longer enough for an IAM vendor to claim they have a powerful AI engine. You must demand that they demonstrate its explainability. Ask to see the explanations generated for different personas. How do they present a risk score to an analyst? How do they justify a blocked transaction to an end-user? A vendor who cannot provide clear, compelling answers to these questions is selling you a black box that you will not be able to govern.

The technical implementation should focus on integrating these explanations seamlessly into existing workflows. An analyst should not have to pivot to a separate "XAI tool" to understand an alert. The explanation should be embedded directly within the security console, right next to the event it is describing. The user should see the simplified justification on the same screen where their access was denied. The goal is to make the explanation a natural and effortless part of the process.

Finally, we must remember that XAI is a tool to augment human intelligence, not to replace it. Even with the best explanations, there will be edge cases, ambiguities, and errors. The ultimate backstop for a flawed machine decision is an empowered human operator. The XAI provides the evidence, but the human remains the judge. A well-designed appeals process, where a user can challenge a machine's decision and have it reviewed by a person, is the final, essential component of a truly accountable system. It is the crucial circuit breaker that ensures our quest for intelligent automation does not lead us to abdicate our own judgment.

* * *

16

Building Access Observability in the IAM 3.0 Era

There is an old parable about three blind men encountering an elephant. One touches the leg and declares it a tree. Another feels the trunk and calls it a snake. The third grabs the tail and insists it is a rope. For decades, this has been a perfect metaphor for investigating an access control problem in a large enterprise. The network team looks at their firewall logs, the identity team checks the single sign-on events, and the application owner stares at their local user database. Each one sees a small piece of the truth, but no one sees the elephant. In the hyper-connected, AI-driven world of IAM 3.0, this fragmented vision is no longer a mere inconvenience; it is a critical security failure.

The systems we have built are now too complex, too distributed, and too dynamic to be understood by looking at one log file at a time. An access decision is no longer a single event occurring on a single server. It is a cascading journey that might begin on a user's mobile device, travel through a cloud-based identity provider, get scored by a machine learning model running in a different cloud, and culminate in a permission being granted or denied by a microservice running in a Kubernetes cluster somewhere else entirely. The traditional tools of monitoring and logging, while still necessary, are like the blind men's hands,they can only feel the part of the problem they are directly touching.

This is the challenge that has given rise to a new discipline: access observability. The term "observability" comes to us from the world of control theory and has been popularized by the site reliability engineering (SRE) community. It is often confused with monitoring, but the distinction is crucial. Monitoring is what you do when you know what you are looking for. You create a dashboard with predefined metrics,CPU usage, login failures per minute, network latency,and you set an alert to fire when a metric crosses a threshold. Monitoring is about watching for known unknowns.

Observability, on the other hand, is about building a system that allows you to explore and understand behaviors you never thought to predict. It is the ability to ask arbitrary questions about the state of your system without having to predefine the question. It is the toolkit for investigating unknown unknowns. If monitoring is having a dashboard of gauges in a car, observability is having the full telemetry data from the engine control unit and a diagnostic computer that lets you ask, "What is the precise fuel injector pressure on cylinder three, correlated with the oxygen sensor readings, over the last 3.7 seconds?"

In the IAM 3.0 era, this capability is not a luxury; it is a necessity. When access is granted by a probabilistic AI, when permissions are ephemeral, and when identities are spread across a dozen different platforms, the old way of troubleshooting is dead. You can no longer ask three different teams to "check their logs" and hope to piece the story together. You need a unified view. You need the ability to see the elephant.

To build a truly observable system, we can borrow the foundational concepts from our SRE colleagues and adapt them to the unique challenges of identity and access. The so-called three pillars of observability are logs, metrics, and traces. In the access control context, they form a powerful hierarchy of insight.

Metrics are the most basic pillar. They are the high-level, aggregated, numerical representation of your IAM system's health. These are the things you put on a dashboard. Examples relevant to IAM include the total number of successful authentications per hour, the rate of MFA challenges, the average risk score generated by your adaptive AI, or the number of just-in-time access grants per day. Metrics are excellent for getting a quick, at-a-glance overview and for alerting on widespread problems. A sudden spike in the 'MFA_failure_rate' metric is a clear signal that something is wrong, perhaps with the MFA service itself or with a phishing campaign in progress. Metrics tell you that a problem exists, but they rarely tell you why.

Logs are the next level of detail. A log is an immutable, time-stamped record of a discrete event. It is the raw, granular evidence. A single successful login might generate dozens of log entries across multiple systems: a log from the web application gateway, an authentication event in the identity provider's system log, an entry in the behavioral analytics engine's log, and an audit event in the target application. Logs are the ground truth. When you need to know exactly what happened at a specific moment in time, you turn to the logs. The problem in a modern environment is not a lack of logs; it is a deluge of them, all stored in different formats, in different locations, with no easy way to connect them.

This brings us to the third and most transformative pillar: traces. A trace is what stitches the disparate logs together into a single, coherent story. In the world of application performance management, a trace follows a single user request as it travels through the various microservices of a distributed application. In access observability, a trace follows a single access decision on its journey. It does this by assigning a unique correlation ID to the request at its inception and ensuring that this ID is passed along at every subsequent step.

Imagine a user, Sarah, clicking a button in a SaaS application to access a sensitive report. A trace would capture this entire lifecycle. It would start with the initial click, which is assigned a 'trace_id'. That 'trace_id' is included in the request sent to the identity provider. The IdP's logs,for credential validation, MFA check, and token issuance,are all tagged with this same 'trace_id'. The signed token, containing the 'trace_id', is sent back to the application. The application calls an AI risk engine to score the request, passing the 'trace_id' along. The AI engine's logs, detailing the signals it used, are tagged. The application then checks its own permissions, logging the outcome with the same 'trace_id'. Finally, access is granted, and the final event is logged.

Now, when a security analyst needs to investigate Sarah's access, they don't have to hunt through five different systems. They can simply query the observability platform for that single 'trace_id'. The result is a beautiful, waterfall-style view showing every step of the process, in order, with the time taken at each step. They can see the entire journey of that one decision, from start to finish. This is the superpower of observability. It turns a multi-hour archaeological dig through fragmented logs into a simple query.

Building this capability requires a deliberate architectural approach. At its heart is a centralized data platform, often called a security data lake, but its function is more active than that of a simple repository. This is not just a place to dump data; it is a processing and analytics engine. The architecture consists of several key layers.

The first layer is Data Collection. The goal is to get data from every system that participates in an access decision. This means deploying agents or configuring data forwarders on everything. This includes your identity providers (like Okta, Azure AD, or Ping), your cloud infrastructure (AWS CloudTrail, Azure Monitor, Google Cloud Audit Logs), your on-premises servers, your network devices, your endpoint security agents, and crucially, your applications themselves. The application developers must become part of the process, instrumenting their code to emit structured logs and propagate the 'trace_id'.

The next layer is the Data Pipeline. As raw data flows in from these myriad sources, it must be processed before it can be stored. This involves parsing the different log formats, normalizing the data into a common schema (so that a 'user_id' from one system means the same thing as a 'principal_id' from another), and enriching the data with additional context. For example, when a log entry with an IP address comes in, the pipeline might enrich it with geolocation data and a threat intelligence reputation score for that IP.

The third layer is Data Storage. The processed data is then loaded into a high-performance database optimized for handling massive volumes of time-series data and executing complex queries at speed. This could be a commercial cloud data warehouse, a specialized observability backend platform, or a custom-built solution. The choice of technology is less important than its ability to ingest data in real time and make it available for querying almost instantly.

Finally, there is the Query and Visualization layer. This is the user interface for your observability platform. It is where analysts, engineers, and even auditors can interact with the data. A powerful query language is essential, allowing users to slice and dice the data in any way they can imagine. This layer also provides the tools to build dashboards for monitoring key metrics and to create visualizations, such as service maps that show how different components of the IAM stack interact, or access graphs that illustrate the complex permission paths between users and resources.

In practice, this observability platform becomes the central nervous system for both security operations and governance. Consider the common problem of debugging a user's access issue. The user complains, "I can't log into the new reporting tool." Without observability, this ticket bounces between the help desk, the identity team, and the application team, with each group claiming their part is working correctly. With observability, the help desk analyst can simply search for the user's name in the observability platform, find the failed login trace, and see the exact error message: 'SAML Assertion Invalid: Audience mismatch.' The trace shows that the identity provider issued a token intended for a different application. The problem is a simple configuration error in the single sign-on setup. A problem that could have taken days to resolve is diagnosed in minutes.

This extends to the most advanced features of IAM 3.0. In the last chapter, we discussed how XAI can explain why an AI model made a particular decision. Access observability provides the context that surrounds that decision. The XAI might tell you that a user's risk score was high because of an impossible travel anomaly. The observability trace would show you the entire event flow: the user's last successful login from San Francisco, the new login attempt from an IP in Belarus, the call to the AI engine, the high risk score being returned, the policy engine blocking the login, and the alert being fired to the security team. It provides the complete, end-to-end narrative.

From a governance perspective, this is a revolution. An auditor no longer has to ask for static, point-in-time reports that are obsolete the moment they are generated. Instead, they can be given read-only access to the observability platform to ask their own questions. An auditor could run a query to find every instance where a user was granted temporary, just-in-time access to a production system and verify that every single grant corresponds to a valid change ticket number. They could query for all users who have access to both create a vendor and approve payments, validating a critical SoD control against the system's actual, real-time behavior. Governance becomes a dynamic, evidence-based process of inquiry rather than a static ritual of attestation.

The journey to building this capability is as much cultural as it is technical. It requires breaking down the silos that have traditionally existed between different IT and security teams. In an observable world, everyone is responsible for instrumenting their systems to provide high-quality data. Application developers cannot treat logging as an afterthought. Cloud engineers must ensure their infrastructure is configured to export the necessary audit data. The IAM team must work to ensure correlation IDs are propagated across the identity lifecycle.

This shared responsibility fosters a shared understanding. When all teams are looking at the same unified set of data, the finger-pointing and blame games that characterize so many incident response efforts begin to fade away. The focus shifts from defending one's own silo to collaboratively solving the problem, using the rich data provided by the observability platform as a common source of truth. It allows you to finally see the whole elephant, in all its complex and magnificent detail.

* * *

17

Sovereign AI and the Global Identity Puzzle

For most of its history, the practice of Identity and Access Management has been blissfully ignorant of geography. It was a problem of logical connections, not physical locations. A user was a string in a directory, an application was an endpoint on a network, and the goal was to securely link the two, regardless of whether they were separated by a firewall or an ocean. The internet itself was conceived as a borderless expanse, a global commons where data flowed freely. This idealized vision, however, is now colliding with the hard realities of geopolitics. The era of the truly global, placeless identity system is ending, and in its place, a far more complex and fragmented landscape is emerging. Nations are reasserting their authority not just over their physical territory, but over their digital territory as well. This is the dawn of digital sovereignty, and its most potent instrument is the concept of Sovereign AI.

Sovereign AI is the simple but revolutionary idea that the artificial intelligence systems making decisions about a nation's citizens, data, or critical infrastructure must be subject to that nation's laws, values, and control. It is a direct rejection of the notion that a single AI, running in a single cloud region owned by a foreign company, should be allowed to govern access or assess risk on a global scale. This principle manifests in several increasingly strict layers, each presenting a new and formidable challenge to the IAM architect. The most basic layer is data residency. Driven by privacy regulations like the European Union's General Data Protection Regulation (GDPR), this mandates that the personal data of a country's citizens must be stored and processed within that country's borders, or within a region with legally recognized equivalent protections. For an IAM system, this means the raw behavioral data, user attributes, and log files used to train an AI model for a German employee must reside on servers within the EU.

The next layer up is model residency. It is no longer enough to just keep the data local; the AI model itself, the trained artifact of code and weights, must also be hosted and executed on servers within the sovereign territory. The logic here is that if the model is running on foreign soil, it could still be subject to foreign laws or intelligence gathering, even if the training data remains local. This prevents a scenario where a model trained on EU data is run in a US data center, where it might be subject to a subpoena under the US CLOUD Act, creating a direct conflict with GDPR.

The final and most profound layer is algorithmic sovereignty. This is the assertion that a nation has the right to dictate the very logic and ethical framework of the AI models making decisions within its purview. The upcoming EU AI Act is a prime example of this thinking. It proposes to classify AI systems by risk level, with "high-risk" systems,a category that would almost certainly include AI used for access to essential services like banking or employment,being subject to stringent requirements for transparency, fairness, and human oversight. A nation might mandate that any AI used for identity verification cannot use certain types of data as input features, or that its decisions must be explainable according to a specific national standard. This moves beyond controlling where the AI runs to controlling how it thinks.

This rise of sovereign AI creates an immediate and intractable puzzle for any organization that operates on a global scale. Consider a large multinational corporation with employees and operations across North America, Europe, and Asia. The CISO's dream is a single, unified IAM platform. They want one adaptive authentication engine, one behavioral analytics platform, and one set of access policies to govern the entire enterprise. It is efficient, consistent, and easier to secure. Sovereign AI shatters this dream into a dozen pieces. That single, global AI engine is no longer legally viable. It cannot legally process data from an EU employee in a US data center, nor can it apply a US-centric risk model to an employee working in China, where data laws are even stricter.

The immediate architectural consequence is that the monolithic global IAM platform is dead. It must be replaced by a federated model of regional, sovereign AI instances. The corporation must now deploy and maintain a "German AI" that runs in Frankfurt, trained on German data, and compliant with the EU AI Act. It needs a "Chinese AI" that runs in Shanghai, respecting China's data localization laws. And it needs an "American AI" for its US operations. These are not just copies of the same model running in different places. They may need to be trained on different data sets and operate under different logical constraints to comply with local regulations. The single, unified brain has been balkanized into a committee of regional minds.

This fragmentation introduces a new and bewildering set of practical challenges, particularly when identities need to cross these new digital borders. Imagine a senior executive from the company's German office travels to the United States for a series of meetings. She opens her laptop in a New York hotel and attempts to log into the corporate network. This single, everyday event triggers a cascade of geopolitical and architectural questions. Which AI gets to make the risk assessment? Is it the German AI, which holds her behavioral baseline but is now seeing a request from a foreign and potentially untrusted jurisdiction? Or is it the American AI, which understands the local network context but knows nothing about this user?

If the German AI is to make the decision, it needs context about the New York network. Can the American AI provide this context? If so, what data can it share without violating any US regulations? And can the German AI process this data without violating GDPR's rules on international data transfers? Conversely, if the American AI takes charge, it needs to know the user's normal behavior. Can the German AI share her baseline? Or is that sensitive personal data that cannot leave the EU? The simple act of a login has become a complex diplomatic negotiation between two sovereign AI instances, each bound by a different set of laws.

Now expand this puzzle beyond the corporate world to the emerging landscape of national digital identities. The EU is moving toward a bloc-wide digital identity wallet under its eIDAS 2.0 framework. This would allow a citizen of any EU member state to use a secure, government-backed digital ID to prove their identity and access services across the union. This is a monumental step forward, but it is still largely confined within the EU's "trust zone." What happens when a French citizen, using their eIDAS wallet, wants to prove their age to rent a car from an American company online? The verification process will inevitably involve AI to assess the risk of the transaction and the authenticity of the presented credential. Whose AI? Governed by which rules? The American rental company's AI will likely be subject to US laws, while the French citizen's identity is a product of EU law. How is trust brokered across this divide?

The legal frameworks codifying this digital nationalism are no longer theoretical. The EU's GDPR has been in effect for years, establishing a mature and heavily enforced regime for data protection and residency. The EU AI Act, once it comes into full force, will layer on top of this, creating a comprehensive regulatory scheme for artificial intelligence. In China, the Cybersecurity Law (CSL) and the Personal Information Protection Law (PIPL) create one of the world's most robust data sovereignty frameworks, making it exceptionally difficult to move data generated within China outside its borders. India, Brazil, and dozens of other nations are following suit, each drafting its own unique set of rules. We are rapidly moving away from a world with a handful of different privacy regulations to a world with dozens, each with its own interpretation of sovereignty.

This creates a compliance nightmare for IAM architects. Building a system that can navigate this patchwork of conflicting regulations requires a new level of sophistication. The IAM platform can no longer be blind to jurisdiction. It must be location-aware at its very core. It needs a policy orchestration layer that can dynamically determine which legal regime applies to any given access request based on the citizenship of the user, their current physical location, and the location of the resource they are trying to access. The system must be able to route the request to the correct sovereign AI engine and ensure that any cross-border data sharing that occurs is permissible under the laws of all involved jurisdictions.

This leads to the concept of "AI trust zones" or geopolitical "digital blocs" We are seeing the emergence of a US-centric digital ecosystem, a European one built around GDPR and the AI Act, and a Chinese one operating under its own distinct principles. Other nations may align with one of these blocs or attempt to create their own. The internet is fracturing, or "splintering," along these geopolitical fault lines. For an IAM system to provide a seamless global experience, it must learn to function as a diplomat and a translator, moving gracefully between these often-suspicious blocs.

The architectural patterns to support this are still emerging, but they will likely revolve around the idea of privacy-preserving communication between these sovereign AI instances. Instead of the German AI asking for the raw network data from the American AI, it might ask a more abstract, yes/no question: "Is this login originating from an IP address on your corporate watchlist?" The American AI can answer this question without revealing the specific IP address or other sensitive network details. This kind of interaction, governed by standardized protocols for inter-AI communication, allows for the sharing of risk signals without the wholesale transfer of raw data. It is a way to build a chain of trust without breaking the laws of sovereignty.

Another critical architectural shift will be the rise of what we might call "context brokers." These would be specialized services that sit at the boundaries between sovereign zones. Their job is not to store data, but to enrich access requests with permissible local context before passing them on. When our German executive logs in from New York, the request might first hit a US-based context broker. This service, operating under US law, could add information like a local threat intelligence score for the hotel's network and then forward the enriched but anonymized request to the German AI for the final decision.

This complex new reality transforms the role of the IAM architect. The job is no longer just about understanding protocols like SAML and OAuth. It now requires a working knowledge of international law, data privacy regulations, and geopolitics. The design of a new authentication flow is not just a technical exercise; it is an exercise in comparative law. Choosing a cloud provider is not just about price and performance; it is a strategic decision about which sovereign trust zone you want to operate in.

The global identity puzzle, therefore, is not a technical problem with a neat technical solution. It is a reflection of a fundamental tension in the modern world: the global, borderless nature of digital technology is running headlong into the timeless, bordered nature of national sovereignty. The result is a messy, complex, and constantly shifting landscape where the rules of identity are being rewritten country by country. The challenge for the architect is to build systems that are not only secure and intelligent, but also worldly and politically astute, capable of navigating a world where an identity's validity can change the moment it crosses a digital border.

***

18

Decentralization Paradox: Blockchain, SSI, and Centralized Models

There are two powerful currents flowing through the world of digital identity, and they are moving in opposite directions. The first is a grassroots, user-centric movement pushing power to the edge, a philosophy that champions individual control, privacy, and consent. It envisions a world where each person is the master of their own identity. The second current is an enterprise-driven, top-down imperative for intelligence and security, a paradigm that requires centralized analysis, correlation, and a holistic view of behavior to make sense of a chaotic threat landscape. One seeks to give the user a veil of privacy; the other seeks to lift all veils in the name of protection. This is the great decentralization paradox, a fundamental tension between how we want our identity to be owned and how we need it to be secured.

At the heart of the first current is the concept of Self-Sovereign Identity, or SSI. It is a radical departure from every identity model that has come before it. For decades, our digital identities have been issued and managed by others. Your corporate identity is owned by your employer, sitting in their Active Directory. Your social identity is owned by a tech giant, residing in their massive user database. You are, in effect, a digital tenant, living in a space controlled by a landlord who can change the locks or evict you at any time. SSI proposes to make you the homeowner. It is a model where individuals create and control their own identities, independent of any central authority.

This model is built on a trio of core technologies. The foundation is the Decentralized Identifier, or DID. A DID is a globally unique, persistent identifier that a user can create for themselves without asking for permission. It is like getting a phone number without needing a phone company. The DID itself doesn't contain any personal information; it is simply a pointer, a way to be found. The public keys needed to interact with that DID are often anchored on a distributed ledger, like a blockchain, which serves as a decentralized and censorship-resistant public directory. This provides a trust anchor, a way for anyone to look up a DID and find the public key needed to verify signatures coming from its owner.

The second piece of the puzzle is the digital wallet. This is not a cryptocurrency wallet (though it can be), but a secure application, typically on a user's phone or computer, that holds their digital identity information. It is the place where the user manages their DIDs and stores their personal data. It is the command center for their self-sovereign life, the digital equivalent of the physical wallet in their pocket.

The final, and most powerful, component is the Verifiable Credential, or VC. A VC is a digital, tamper-proof version of the credentials we use every day: a driver's license, a university diploma, an employee badge, or a concert ticket. It is a set of claims about a subject, issued by a trusted entity (the issuer), and held by the subject (the holder) in their digital wallet. For example, a university (the issuer) could issue a VC to a student (the holder) that contains claims like "Graduated with a B.S. in Computer Science" and "Conferred on May 15, 2024." This credential is digitally signed by the university using a key linked to their public DID.

The magic of VCs lies in how they are used. The holder can present this credential to a third party (the verifier) who needs to confirm their degree. The verifier can instantly check the digital signature against the university's public DID on the blockchain, proving both the credential's authenticity and its integrity without ever having to contact the university directly. This creates a "trust triangle" that is efficient, secure, and decentralized. Furthermore, VCs enable selective disclosure. The student could choose to prove only that they have a degree from that university without revealing the specific degree or graduation date, giving them granular control over their own data. This is the promise of SSI: a more private, secure, and user-controlled internet.

This vision, however, runs headlong into the reality of IAM 3.0. As we have seen, the intelligence that powers modern access control is predicated on data aggregation. The probabilistic risk engines, the behavioral analytics platforms, and the autonomous identity systems at the heart of the new paradigm are voracious. They crave context. Their accuracy is directly proportional to the volume and variety of data they can ingest and correlate. A behavioral model cannot establish a baseline for a user by observing a single, isolated transaction. It needs to see all of them: the logins, the application usage, the data access patterns, the network locations. The more data it can correlate, the smarter it becomes.

This is the core of the paradox. SSI is a system explicitly designed to prevent the unauthorized correlation of a user's activity across different contexts. The user's wallet is supposed to be a black box, revealing only what the user consents to reveal, for a specific purpose, at a specific time. An AI-driven IAM system, on the other hand, is designed to be an all-seeing eye, constantly correlating activity to build a holistic picture of risk. One system is built on a principle of data minimization and disaggregation; the other is built on a principle of data maximization and aggregation. They are fundamentally oil and water.

Let's drill down into the practical points of conflict. Consider the behavioral analytics engine, a cornerstone of adaptive access. Its job is to learn the unique rhythm of a user's digital life. It learns that Bob, an accountant, typically logs in from the Chicago office between 8 AM and 6 PM, uses the ERP system and a spreadsheet program, and rarely accesses more than 100 financial records a day. Now, imagine Bob is operating in a pure SSI world. To access the ERP system, he presents a verifiable credential from his wallet,an "employee badge" VC issued by HR. To access the accounting software, he presents the same VC. The system sees two separate, verifiable presentations. But the SSI philosophy is designed to make it difficult for the ERP system and the accounting software to collude and realize they are talking to the same "Bob." If his DIDs are pairwise and unique for each interaction, how can the centralized AI engine link these events to build a single behavioral baseline for "Bob"? The very privacy feature that protects the user cripples the intelligence of the security system.

The conflict is just as stark when it comes to risk scoring. Let's say a user wants to access an online service that requires them to be over 18. Using a VC, they can prove this fact without revealing their name, address, or exact date of birth. This is a huge win for privacy. But the service's adaptive AI engine might have a much more demanding view of risk. Its model might have learned that for this particular transaction, users between the ages of 18 and 25 represent a higher fraud risk. The model wants the exact date of birth, not just a "yes/no" answer to the over-18 question, so it can factor that into its probabilistic score. The user's desire for selective disclosure is in direct opposition to the model's need for rich, contextual features to make what it considers to be an accurate judgment.

The autonomous identity systems discussed in previous chapters represent an even more profound philosophical clash. An autonomous engine works by constantly observing the user's digital environment. It sees the project ticket assigned to a developer, reads their team chats to infer context, and analyzes the access patterns of their peers to proactively grant them the permissions they will need. This requires a level of persistent, background surveillance that is the antithesis of the SSI model, which is based on discrete, user-initiated, consent-driven interactions. The autonomous system wants to be a prescient butler who anticipates your every need; the SSI model insists that the butler must knock and ask for permission before entering any room.

The role of the blockchain in this architecture is often misunderstood, adding another layer of complexity to the paradox. The common misconception is that personal data is stored on the blockchain. In a well-designed SSI system, this is never the case. A blockchain is a public, immutable database, making it the worst possible place to store sensitive information. Instead, the blockchain is typically used only to anchor the DIDs and the public keys of the issuers. It is the public bulletin board where you can go to verify that a given DID is real and find the key needed to check a signature. The Verifiable Credentials themselves, containing the actual personal data, live off-chain, held securely in the user's private wallet.

However, even this limited use of a public ledger creates governance challenges. While no personal data is on the chain, the metadata of the transactions,which DID interacted with which verifier at what time,can be recorded. This trail of public metadata can, in itself, leak information. Furthermore, the immutability of the blockchain runs into direct conflict with privacy regulations like GDPR, which includes a "right to erasure" or "right to be forgotten." How can you erase a user's DID from a ledger that is, by its very nature, designed to be permanent and unchangeable? These are difficult questions that architects and legal teams are still grappling with.

This fundamental paradox leaves the IAM architect in a seemingly impossible position. The business demands the intelligence and efficiency of an AI-driven, centralized risk engine. Simultaneously, a growing chorus of users, regulators, and privacy advocates is demanding the control and consent offered by a decentralized model. You are being asked to build a system that is simultaneously open and closed, transparent and opaque, aggregated and disaggregated. You are being asked to build a square circle.

The result is that organizations are forced to consider complex, hybrid architectures that attempt to bridge this chasm. These are not clean, elegant solutions but a series of compromises and clever workarounds designed to satisfy two masters who want completely different things. It is clear that the existing architectural patterns are insufficient. We cannot simply plug a decentralized wallet into a centralized AI and expect them to work in harmony. The philosophical divide is too great.

One potential path forward involves creating new architectural components that can mediate between these two worlds. Perhaps a system could be designed where the user's wallet itself contains a tiny, personalized machine learning model. This "edge AI" could learn the user's baseline locally, on their own device, without ever sending the raw activity data to a central server. The wallet could then generate a "proof of normalcy" credential, a VC that simply attests, "My behavior over the last 24 hours has been within 99% of my established baseline." A central risk engine could consume this privacy-preserving proof as one of its input signals, allowing it to benefit from behavioral analysis without ever seeing the behavior itself.

Another approach centers on the concept of reputation. Instead of a centralized system tracking a user's every move, the user's wallet could accumulate reputation VCs from various interactions. A financial institution could issue a credential attesting that the user has a five-year history of on-time payments. An e-commerce site could issue one for a hundred successful transactions with no chargebacks. The user could then present a collection of these reputation credentials, which together would paint a picture of trustworthiness that the central risk engine could use as a proxy for a detailed behavioral profile.

These are early, emerging ideas, and they are not without their own challenges in terms of complexity, cost, and standardization. They highlight the fact that resolving the decentralization paradox is not about choosing one model over the other. It is not a battle where decentralization will "win" or centralization will "win." The future of identity will inevitably be a hybrid, a messy and negotiated settlement between these two opposing forces. It requires a new layer of architecture, a new kind of intelligent middleware, that can sit between the user's sovereign domain and the enterprise's security perimeter. This new component must act as a translator, a diplomat, and a broker of trust, capable of satisfying the user's demand for privacy while still providing the enterprise with the signals it needs to stay secure. This is the domain of the Trust Mediation Engine.

* * *

19

Trust Mediation Engines: The New Architecture Layer

We have arrived at an impasse, a place where two of the most powerful and promising trends in digital identity are locked in a philosophical stalemate. On one side stands the decentralized movement, a user-centric crusade for privacy and control, championed by technologies like Self-Sovereign Identity and Verifiable Credentials. On the other stands the centralized intelligence of IAM 3.0, an enterprise-driven necessity that relies on data aggregation and holistic analysis to defend against advanced threats. One seeks to build walls around user data; the other needs to look over those walls to see the big picture. An architecture that attempts to serve both of these masters is destined for failure, unless we introduce a new component designed specifically to negotiate a peace between them. This is the role of the Trust Mediation Engine (TME), a new architectural layer that functions as a diplomat, a translator, and a security checkpoint between the sovereign territory of the user and the protected domain of the enterprise.

A Trust Mediation Engine is not just another policy engine or API gateway. It is a specialized, intelligent middleware built to solve the decentralization paradox. It is designed to sit at the very edge of the enterprise trust boundary, acting as the primary point of contact for interactions originating from decentralized identity wallets. Its core purpose is to receive and understand the privacy-preserving credentials presented by a user, enrich them with context that only the enterprise can know, and transform them into a set of signals that a centralized AI risk engine can consume, all without fundamentally violating the user's control over their own data. If SSI is about creating different languages of identity for different contexts, the TME is the universal translator that allows them to be understood by the enterprise's native tongue of risk assessment.

The first and most fundamental function of a TME is credential validation and interpretation. When a user presents a Verifiable Credential (VC) from their digital wallet, the TME's first job is to act as the verifier. It must perform the essential cryptographic checks: confirming that the credential has not been tampered with and validating the issuer's digital signature. This process typically involves looking up the issuer's Decentralized Identifier (DID) on a public, distributed ledger like a blockchain to retrieve their public key. This establishes the basic authenticity of the credential, it proves that the digital driver's license was, in fact, signed by the Department of Motor Vehicles. But this is just table stakes. A TME must go beyond simple validation and begin to interpret the credential in the context of the requested transaction. It must understand the semantics of the claims, not just the syntax of the data.

This leads to the TME's most critical and sophisticated function: privacy-preserving data transformation. This is the heart of its diplomatic mission. A centralized AI risk engine, as we know, thrives on rich, detailed data. A user's SSI wallet, by design, wants to provide the absolute minimum data necessary. The TME must bridge this gap. Imagine a user applying for an online loan. The lender's risk model wants to know the applicant's exact credit score. The user, however, may only want to present a VC from a credit bureau that attests, "Credit score is above 750." The TME receives this binary credential. Instead of just passing a "yes" to the risk engine, it transforms it. It might convert the credential into a standardized signal, like 'credit_worthiness_tier: 'prime''. This gives the AI a usable feature for its model without exposing the user's exact score. The TME acts as a data abstraction layer, converting specific, private claims into generalized, risk-relevant signals.

This transformation becomes even more important for behavioral signals. As we discussed, a centralized behavioral analytics engine is crippled if it cannot link user activities together. A TME can help solve this through the use of more advanced, privacy-preserving credentials. A user's wallet might contain a local "edge AI" that learns their behavior on their own device. This local model could generate and sign a VC that attests, "The holder's activity in the last hour was consistent with their established 90-day baseline." The user presents this "proof of normalcy" credential to the TME. The TME validates the credential and passes a single, powerful signal to the central risk engine: 'behavioral_anomaly: 'false''. The enterprise learns what it needs to know, that the user's behavior is normal, without ever seeing the raw, underlying activity data that generated the proof.

The TME also functions as a context enrichment point. A user's wallet knows a lot about the user, but it knows very little about the context of the transaction from the enterprise's perspective. The TME sits in a privileged position where it can see both sides. As it receives the user's VC, it can simultaneously query internal enterprise systems to gather additional risk signals. It can take the source IP address from the incoming request and check it against an internal threat intelligence feed. It can trigger a request to a device management system to get a real-time posture assessment of the user's endpoint. It can look up the resource being requested and attach a "data sensitivity" tag. The TME then aggregates these enterprise-side signals with the user-provided signals from their VCs, creating a much richer, 360-degree view of the transaction for the AI risk engine.

In this role, the TME becomes a critical policy enforcement point that operates before the core AI engine is even engaged. An organization can configure policies directly on the TME to act as a first line of defense. A policy might state that any VCs used for identity verification must have been issued by one of a pre-approved list of identity proofing services. If a user presents a credential from an unknown or untrusted issuer, the TME can reject the transaction immediately, without ever bothering the downstream systems. Another policy could define the required "level of assurance" for different transactions. Logging in to view a marketing site might only require a simple "employee" VC, but accessing the source code repository might require the presentation of both an "employee" VC and a "completed security training" VC. The TME enforces this credential choreography, ensuring the user has presented the right combination of proofs before the request is allowed to proceed.

Architecturally, the TME sits at a critical junction. It is the new front door for identity verification in a decentralized world. It lives at the edge of the corporate network, exposed to the public internet to receive requests from user wallets. From this position, it communicates outwards to public trust infrastructure, like blockchains or other verifiable data registries, to validate DIDs. It communicates inwards to the core IAM stack, feeding its transformed and enriched signals into the adaptive authentication and risk engines. And it communicates sideways to other enterprise security services, like threat intelligence platforms and endpoint management systems, to gather contextual data. This placement makes it a powerful control point, but also a very attractive target that must be secured with the utmost rigor.

Let's walk through a practical example of the TME in action. A member of a corporate supply chain, an employee of a partner company, needs to access a shared logistics portal. In the old world, the corporation would have to create and manage an account for this partner. In the SSI world, the partner uses their own identity.

1. Presentation: The partner employee, using their corporate digital wallet, presents a VC to the portal's login page. This VC, issued by their

own employer, attests to their name, role ('Logistics Coordinator'), and employment status.

2. Mediation (The TME's Role):

Validation: The portal forwards the VC to the enterprise TME. The TME looks up the partner company's public DID, retrieves their key, and verifies the signature on the VC. It confirms the credential is authentic.

Policy Enforcement: The TME checks its policies and confirms that this partner company is on its list of trusted suppliers and that a "Logistics Coordinator" role is permitted to access the portal.

Context Enrichment: The TME logs the source IP of the request and sees that it's coming from a network not previously associated with this partner. It adds a low-level risk flag.

Transformation: The TME transforms the specific claims from the VC into generalized signals: 'user_role: 'partner_logistics'', 'employment_status: 'active'', 'issuer_trust_level: 'high''. It adds its own signal: 'network_anomaly: 'true''.

3. Risk Assessment: The TME passes this bundle of clean, privacy-preserving signals to the central AI risk engine.

4. Decision: The AI engine evaluates these signals. The high trust level of the issuer and the valid role outweigh the minor network anomaly. It returns a low risk score. The TME receives this score and instructs the portal to grant the user access.

This entire interaction happened without the enterprise ever needing to create an account for the partner or see any more of their personal data than was absolutely necessary. The TME brokered the trust, translating the decentralized credential into the language of centralized risk.

This mediating role also positions the TME as a key player in solving the sovereign AI puzzle. As we've discussed, data residency and algorithmic sovereignty are fracturing the global identity landscape. A TME can be deployed at the digital border of each sovereign zone. A German TME, operating under EU law, could be responsible for all interactions with

EU eIDAS wallets. When a French citizen attempts to access a US-based service, the request would first hit the US company's American TME. This TME, recognizing the credential's origin, could then initiate a secure, policy-governed dialogue with the user's home TME in France. This inter-TME communication would negotiate the exchange of the minimum necessary proof, respecting the legal constraints of both jurisdictions. The TME becomes the diplomatic attach√© that allows for cross-border trust without violating digital sovereignty.

Of course, the path to implementing this vision is not without its own set of significant hurdles. The first is a lack of mature standards. For TMEs to be interoperable, there must be common, agreed-upon protocols for everything from how a TME discovers a wallet's capabilities to the format of a privacy-preserving risk signal. These standards are still in their infancy. The second challenge is complexity. This architecture introduces another sophisticated, stateful component that must be designed, built, secured, and maintained. It is not a simple, off-the-shelf product category today.

Finally, the TME itself becomes a critical point of trust and a high-value target. An organization's entire decentralized identity strategy rests on the security and integrity of its TME. If an attacker can compromise the TME, they could potentially forge validated credentials, suppress risk signals, or bypass policies for the entire enterprise. The governance of the TME,who defines its rules, who can alter its transformation logic, and how its own actions are audited,becomes a new and critical discipline for the identity team.

Despite these challenges, the concept of the Trust Mediation Engine provides a compelling and logical path forward. It acknowledges the immovable reality of the decentralization paradox and offers a pragmatic design pattern for resolving it. It accepts that the future of identity is not a choice between a fully centralized or a fully decentralized model, but a hybrid that requires a new and intelligent layer to bridge the two worlds. Building this bridge will be one of the great architectural challenges of the next decade, but it is the essential work required to build an identity ecosystem that is simultaneously intelligent, secure, and respectful of the individual.

* * *

20

Risk Dynamics in Smart Identity Systems

There is a strange and unsettling paradox at the heart of modern enterprise security. Organizations are pouring unprecedented resources into smarter, more intelligent Identity and Access Management systems. We are replacing rigid rules with adaptive AI, deploying behavioral analytics that can spot a pin-drop of anomalous activity in a sea of data, and building automated platforms that promise a future of frictionless, prescient security. And yet, if you survey the landscape of CISOs and security practitioners, you will not find a uniform sense of calm. In many corners, the anxiety has only deepened. The very tools that promise to make us safer are subtly changing the nature of risk itself, making it more complex, more insidious, and harder to grasp.

The fundamental bargain of IAM 3.0 is that we are trading simple, known risks for complex, systemic ones. The old world of IAM was fraught with peril, but the dangers were, for the most part, straightforward. A weak password could be guessed. A misconfigured user group could grant excessive privilege. A stolen credential gave an attacker a key to a specific door. The risks were discrete, localized, and followed a relatively linear logic. The new world is different. We have not eliminated risk; we have transformed it. We have exchanged the risk of a broken lock for the risk of a sentient security guard who can be deceived, subverted, or brainwashed. Understanding this new dynamic is crucial to surviving it.

In the legacy model, risk was largely contained. A compromised server was a problem for that server and the applications running on it. A stolen user account was a problem for that user and the handful of systems they could access. The blast radius of a single failure was, while potentially serious, often limited by the siloed nature of the IT environment. This containment has been systematically dismantled by the very technologies that drive modern business. In our hyper-connected, federated world, risk is now a distributed phenomenon, a contagion that can spread through the ecosystem with terrifying speed and efficiency.

The very concept of federation, which allows a user to log in once to access dozens of different cloud services, creates a network of implicit trust. This trust is the superhighway upon which compromise now travels. A single compromised identity provider, the central anchor of the federated system, can become a skeleton key that unlocks every connected application. An attacker who forges a SAML token or a JWT doesn't just get access to one system; they get access to all of them, almost simultaneously. The risk is no longer localized to one application's user database; it is distributed across the entire trust fabric of the enterprise.

This distribution is further amplified by the explosion of non-human identities. The attack surface is no longer just the collection of laptops and phones used by your employees. It now includes a sprawling, poorly governed population of API keys, service accounts, IoT devices, RPA bots, and embedded AI agents. Each of these entities has an identity, requires access, and represents a potential point of failure. Governing this vast and diverse population with tools designed for human users is like trying to manage a city's traffic with the rules of a high school hallway. The scale and behavior are completely different, and the risk they introduce is diffuse and incredibly difficult to track. A single, forgotten API key with excessive permissions, buried in the configuration file of a long-deprecated application, can be a silent, open door into the heart of the network.

Alongside this distribution of risk, adversaries have acquired a new and powerful tool: the invisibility cloak. The cat-and-mouse game between attacker and defender has always been one of stealth versus detection, but AI has given the adversary a decisive advantage in the art of hiding. The days of easily spotting a phishing email because of its poor grammar and clumsy formatting are over. Generative AI can now craft flawless, personalized, and contextually aware social engineering lures at an industrial scale. It can scrape LinkedIn to find a target's new manager and send a perfectly worded "welcome" email containing a malicious link to a fake HR document. This isn't just a better lure; it's a lure that is functionally indistinguishable from legitimate communication, making human error almost inevitable.

This new invisibility extends beyond the initial point of entry. Once inside, a sophisticated attacker no longer behaves like a bull in a china shop. They know our defenses are now tuned to look for jarring anomalies. So, they practice a form of digital method acting. Using the compromised credentials of a real user, they will patiently learn and mimic their victim's established behavioral baseline. They will log in at the right times, from plausible locations, using the correct applications, and accessing data in a way that is consistent with the user's normal patterns. This "low and slow" approach is designed to fly completely under the radar of the behavioral analytics engine, effectively using our own intelligent defenses as a smoke screen for their activity.

The most advanced adversaries are taking this a step further, using AI to create legitimate-looking identities from whole cloth. These synthetic identities, pieced together from fragments of real and fabricated data, can be used to pass automated KYC (Know Your Customer) and identity-proofing checks. The attacker can create a "real" customer account, a "valid" new employee, or a "legitimate" third-party vendor that is, in fact, a sleeper agent.

This entity can then operate within the system for months, building up a history of normal activity, before being activated to perpetrate fraud or steal data. The risk is invisible because, from the system's perspective, there is no intrusion; the threat was invited in through the front door.

Perhaps the most profound change in risk dynamics is the shift from localized failures to systemic fragility. The danger is no longer just a breach of a specific application but a catastrophic failure of the entire underlying trust infrastructure. This introduces a class of risk that is almost existential in nature, threatening to invalidate the foundational assumptions upon which our security is built.

We have already explored the systemic fragility introduced by quantum computing. The "Harvest Now, Decrypt Later" strategy creates a latent risk in every piece of encrypted data that leaves our networks. It means that the security of our most sensitive, long-term data is not a current state but a future bet, a bet that our migration to post-quantum cryptography will win the race against the development of a cryptographically relevant quantum computer. A loss in that race would represent a systemic failure of confidentiality on a global scale.

The AI models at the heart of IAM 3.0 introduce their own form of systemic fragility. In the old world, a misconfigured rule affected only the access governed by that rule. In the new world, a single compromised AI model can undermine the integrity of every access decision across the entire enterprise. An AI poisoning attack that teaches a behavioral model to trust a malicious IP address doesn't just create one backdoor; it gives that IP address a free pass for every user and every application. The decision-making engine for the company has been corrupted. The failure is not in one component but in the central brain.

Even the very intelligence we program into these systems can become a source of systemic risk. Algorithmic bias, as we have seen, is not just an ethical concern; it is a serious operational and compliance risk. A model that learns to unfairly penalize a certain class of users can lead to cascading failures: legitimate users being locked out of critical systems, degradation of the user experience leading to shadow IT, and massive legal exposure for discriminatory practices. The flaw is not a bug in the code that can be patched; it is a fundamental error in the system's learned logic, and its effects are felt system-wide, with every prediction it makes.

Compounding all of these changes is the raw acceleration of compromise. The timeline for a security breach has been brutally compressed. The classic attack lifecycle,weeks of quiet reconnaissance, followed by months of slow lateral movement,was a product of human speed and human limitations. An AI-driven attacker operates on a different clock. Automated systems can scan millions of potential targets, identify a vulnerability, exploit it, and exfiltrate data in a matter of hours, or even minutes. By the time a human analyst sees the first alert, the attacker's objectives may already have been achieved.

This machine-speed risk is amplified by the interconnectedness of our systems. A single compromised session token can be used to pivot across a dozen federated cloud services in the blink of an eye. An automated script can use a stolen API key to make thousands of unauthorized calls before rate limiting or anomaly detection can even kick in. The speed of the attack now routinely outpaces the speed of human-led response. This forces a difficult conclusion: the only way to effectively respond to a machine-speed threat is with a machine-speed defense. This means incident response itself must become more automated, more intelligent, and more integrated into the IAM fabric.

So, with all this in mind, is the risk actually growing, or are we simply more aware of the complexity that was always there? The honest answer is both. There is no question that our instrumentation is better than it has ever been. Tools for access observability have given us the ability to see the "elephant," to trace the journey of an access decision across a dozen different systems. We have a far more detailed and sophisticated map of the risk landscape than we did a decade ago. It is tempting to conclude that the territory just looks more dangerous because the map is better.

But this view is incomplete. Awareness does not create risk; it reveals it. And what our new, higher-fidelity map is revealing is a territory that has genuinely become more treacherous. The nature of the risk itself has changed. We have traded the predictable danger of a crumbling stone wall for the unpredictable danger of a dense, fog-filled jungle. The jungle is populated with camouflaged predators who can mimic their surroundings, where a single misstep can lead to a systemic chain reaction, and where the ground itself is tectonically unstable.

Our improved vision is essential for navigating this new terrain, but we must not mistake seeing the danger for being safe from it. The risks are not just an illusion created by better tools. They are a real and direct consequence of the complex, intelligent, and deeply interconnected systems we have built. The challenge now is to use our newfound awareness not just to watch the risks, but to fundamentally redesign our systems to be resilient to them. The goal can no longer be to build an impenetrable fortress. The fortress is an illusion. The goal must be to build a resilient ecosystem that can bend without breaking, that can detect and isolate failures quickly, and that can continue to function even in the face of partial compromise.

* * *

21

Identity Federation, Passwordless, and Continuous Access

The old model of digital access was a story told in two distinct parts: a moment of entry and a period of presence. You arrived at the castle gate, presented your password as a secret passphrase, and once inside, you were generally considered a trusted resident, free to roam within the courtyards you were permitted until your time was up. This binary, one-and-done approach to authentication was the bedrock of security for decades. It was also, as we have seen, fundamentally flawed. In the dynamic, perimeter-less world of IAM 3.0, trust cannot be a one-time transaction. It must be a living, breathing state that is continuously earned, meticulously verified, and constantly re-evaluated.

This shift from a static event to a dynamic state is not a single technological change but a symphony of three interlocking movements. The first is identity federation, the evolution of single sign-on from a simple convenience into the essential trust fabric connecting a sprawling ecosystem of applications. The second is the passwordless revolution, a fundamental reimagining of the act of authentication that breaks the cycle of weak, shareable secrets. The third, and perhaps most crucial, is the principle of continuous access, a Zero Trust philosophy made real, where trust is never assumed and is subject to scrutiny with every click and every API call. Together, these three components form the active, operational core of a modern identity architecture, turning the theoretical promise of probabilistic IAM into a practical, resilient defense.

Identity federation is not a new concept. Standards like Security Assertion Markup Language (SAML) and OpenID Connect (OIDC) have been the workhorses of corporate single sign-on (SSO) for years, liberating users from the tyranny of remembering dozens of different passwords. In the context of IAM 2.0, federation was primarily a story about user convenience and centralized password policy. It allowed an enterprise to act as a central Identity Provider (IdP), asserting the identity of its users to a host of third-party applications, or Service Providers (SPs). A user would log in once to their corporate IdP, which would then issue a cryptographically signed token,a SAML assertion or a JSON Web Token (JWT),that served as a temporary, portable passport, allowing them to enter other applications without being challenged again.

In the IAM 3.0 paradigm, the role of federation has evolved dramatically. It is no longer just the plumbing for convenience; it is the high-speed data bus for trust signals. The passport it issues is no longer a simple document with a name and a picture. It is now a rich, multi-page dossier, packed with the contextual information that downstream applications and risk engines need to make intelligent access decisions. The modern JWT or SAML assertion is a vessel, and its cargo has become far more valuable than just a username.

When a user authenticates to a sophisticated, IAM 3.0-aware IdP, the token that gets minted can include a wealth of information derived from the adaptive authentication process. It can contain a claim for the Authentication Method Class Reference (AMCR), indicating not just that the user authenticated, but how they did it. Did they use a simple password, a phishing-resistant FIDO2 key, or a one-time code sent via SMS? Each of these carries a different weight of assurance. The token can carry a claim representing the confidence level or risk score calculated by the AI engine at the moment of login. It can include signals about the device's posture, its health, its ownership, and its compliance status.

This enriched token is then presented to the Service Provider. The application is no longer just making a binary decision based on the user's identity. It can now make a much more nuanced, risk-informed authorization decision. An application might have a policy that allows any authenticated user to view its main page, but to access the administrative functions, the token must contain a claim indicating that the user authenticated with a phishing-resistant method and that the login risk score was below a certain threshold. Federation, in this model, is the mechanism that transports the verdict of the central AI brain to all the remote limbs of the enterprise body.

This creates an intricate web of trust relationships that can become incredibly complex. In a multi-cloud, multi-vendor world, it is common for federations to be chained together. A user might authenticate to their corporate IdP, which then asserts their identity to a major cloud provider's IAM system, which in turn grants them access to a specific application running in that cloud. This multi-hop journey can dilute the original trust signal. Keeping track of the entire chain of trust, from the initial authentication to the final access decision, requires the kind of end-to-end visibility provided by the access observability traces we discussed in Chapter Sixteen. Without a unified trace, debugging a federated login failure can feel like trying to find a dropped package somewhere along a thousand-mile shipping route.

For all its importance, federation only solves half the problem. It provides the rails for identity to travel, but it does not, by itself, fix the inherent weakness of the initial login. As long as that first step relies on a password, the entire trust chain is built on a foundation of sand. This is where the second major pillar, the passwordless movement, comes into play. The case against passwords has been made so many times it hardly needs repeating. They are forgotten by users, stolen by attackers, phished with ease, and represent the single greatest source of friction and failure in the user authentication experience. Passwords are a 1960s technology that has long outlived its usefulness.

The goal of passwordless authentication is to completely remove shareable secrets from the process. Instead of relying on something the user knows (a password), it relies on something they have (a registered device) and something they are or do (a biometric or a local gesture). The gold standard for this today is the set of protocols championed by the FIDO Alliance, particularly FIDO2. When a user registers for a service using FIDO2, their device,be it a laptop with a fingerprint reader, a phone with facial recognition, or a physical USB security key,generates a unique public-private key pair for that specific service. The private key is stored securely on the device, often in a specialized hardware-backed secure element, and it never leaves. The public key is sent to the service and registered with the user's account.

When the user wants to log in, the service sends a challenge to the browser or operating system. The user is prompted to unlock their private key using a simple, local action,touching a fingerprint sensor, looking at their phone, or entering a local PIN for their security key. Once unlocked, the private key is used to cryptographically sign the challenge, and this signed response is sent back to the service. The service verifies the signature using the user's registered public key, and if it is valid, the login is successful.

The security of this model is transformative. There is no password for a phishing site to steal. There is no shared secret on the server for an attacker to compromise in a database breach. The credential is bound to the specific origin of the service, meaning a key registered for <code>mybank.com</code> cannot be used to log into a fake site at <code>mybank.attacker.com</code>. It effectively short-circuits the vast majority of credential-based attacks.

In the context of IAM 3.0, a successful FIDO2 authentication is one of the highest-quality signals an adaptive risk engine can receive. It proves, with cryptographic certainty, that the user is in possession of a specific device that they have previously registered. This is a far more reliable indicator of legitimacy than simply knowing a secret that could have been stolen from anywhere in the world. When a risk engine sees a FIDO2 authentication, it can start the user's session with a very high level of baseline trust, which in turn leads to a more frictionless experience. Because the initial authentication is so strong, the system can be less paranoid about subsequent actions, reserving its step-up challenges for only the most genuinely anomalous or high-risk activities. Passwordless authentication doesn't just make the login more secure; it makes the entire adaptive security model work better.

This brings us to the third pillar, the one that ties everything together into a coherent Zero Trust strategy: continuous access. The philosophy here is that authentication is not a single point-in-time event. Even a strong, passwordless login only proves that the user was legitimate at the exact moment they authenticated. What happens five minutes later? Ten minutes later? Has the context changed? Has the device's security posture degraded? Has the user's behavior deviated from the norm? Continuous access is the practice of constantly asking and answering these questions throughout the entire lifecycle of a user's session.

This is not about repeatedly forcing a user to log in. That would be an operational nightmare. Instead, it is about the silent, continuous evaluation of risk signals behind the scenes. The IAM system, fueled by the data flowing into its observability platform, acts as a vigilant chaperone. It watches the session, comparing the ongoing activity against the user's established behavioral baseline, monitoring for changes in device health, and evaluating the sensitivity of the resources being accessed. For 99% of the time, this monitoring is completely invisible to the user. They go about their work, blissfully unaware of the constant risk assessment happening in the background.

It is only when the system detects a significant change in the risk calculus that it intervenes. The trust score, which started high after a strong passwordless login, might begin to drop. Perhaps the user, whose session started on a secure corporate network, has now moved to an untrusted public Wi-Fi hotspot. Perhaps an endpoint security agent reports that a piece of malware has been detected on their device. Perhaps the user is attempting to access a highly privileged administrative function for the very first time.

When the trust score crosses a predefined policy threshold, the continuous access system springs into action with a response that is proportional to the perceived risk. This is not a simple allow/deny decision. It is a spectrum of possible interventions. For a minor increase in risk, the most common response is a step-up authentication challenge. The system will pause the user's action and prompt them to re-verify their identity, perhaps by tapping their fingerprint sensor again. It is a gentle but firm "are you still there, and are you still you?" check.

For a more significant drop in trust, the system might employ session scoping. It can dynamically reduce the user's permissions in real-time, without terminating their session. A user who was logged in with administrative rights might suddenly find themselves demoted to a read-only user. This allows the system to contain the potential blast radius of a session that might have been hijacked, while still allowing the legitimate user to continue with low-risk tasks.

In high-risk scenarios, such as when a user on an unmanaged device is attempting to access a system known to be a frequent target of attacks, the system might invoke session isolation. The user's connection is transparently rerouted through a remote browser isolation (RBI) platform. The application is rendered in a secure, disposable container in the cloud, and only a safe stream of pixels is sent to the user's device. The user can interact with the application, but no active code from the web page ever reaches their endpoint, neutralizing the threat of drive-by downloads or other web-based malware.

Finally, in the face of a critical risk event,such as the impossible travel scenario we've discussed previously,the response is swift and decisive: session termination. The system immediately invalidates the user's session token, logs them out of all applications, and may even temporarily suspend their account to prevent any further attempts. The connection is severed, and an automated alert is fired to the security operations center.

The true power of this model is realized when all three pillars work in concert. Imagine a doctor in a hospital. She starts her day by tapping her hospital-issued ID card to a reader, which triggers a passwordless login to her workstation using a FIDO-compliant credential stored on the card. This high-assurance login is passed via federation to the hospital's Electronic Health Record (EHR) system, granting her standard access. Later, she walks into a patient's room and taps her card on the bedside terminal. The continuous access system detects this as a low-risk continuation of her existing session and seamlessly logs her in. However, when she attempts to access the patient's sensitive psychiatric notes, the system recognizes this as a high-risk action. It triggers a step-up challenge, prompting her to provide a fingerprint on the terminal's built-in sensor before revealing the sensitive data. After she leaves the room, her session on the terminal is automatically terminated. This is IAM 3.0 in practice: secure, adaptive, context-aware, and built around the natural workflow of the user.

Of course, the journey to this state is not without its challenges. There are legacy applications that do not speak modern federation protocols and require cumbersome gateways or agents. User adoption of passwordless methods, while generally positive, requires education and a thoughtful rollout strategy. And tuning a continuous access system is a delicate art; it must be sensitive enough to catch real threats without becoming so "chatty" that it creates a constant stream of disruptive challenges for legitimate users, leading to fatigue and frustration. Navigating these practical hurdles requires careful planning, but the architectural destination is clear. The combination of a robust federation fabric, strong passwordless authentication as a starting point, and vigilant continuous access evaluation throughout the journey is no longer a futuristic vision. It is the essential, foundational grammar of trust in the modern enterprise.

* * *

22

Emerging Attack Surfaces: Bots, APIs, and Synthetic Identities

The traditional landscape of identity and access management was, for a long time, a reassuringly human-centric world. The primary entities we worried about were people: employees, contractors, partners, and customers. Our security models, our governance frameworks, and our mental maps were all built around the concept of a human user sitting at a keyboard, attempting to access a resource. This picture, while still relevant, is now dangerously incomplete. Our digital ecosystems are no longer populated solely by humans. They are teeming with a vast and rapidly growing population of non-human workers and ephemeral actors, each with its own identity, its own access needs, and its own unique potential for compromise. This new workforce of bots, services, and fabricated personas represents a new and shadowy attack surface, one that our human-centric security tools often struggle to see, let alone govern.

The most visible members of this new digital workforce are Robotic Process Automation (RPA) bots. These are not the sentient androids of science fiction but sophisticated software scripts designed to mimic human users to automate repetitive, rules-based tasks. A bot might be programmed to log into an email inbox, open attachments containing invoices, scrape the relevant data, and enter it into a financial application. For the business, this is a miracle of efficiency, freeing up human employees from mind-numbing work. For the identity architect, it is a new class of privileged user that operates 24/7, works at machine speed, and has no manager to call if it goes rogue.

The identity and access challenges posed by RPA are immediate and profound. How does a bot authenticate? In far too many real-world deployments, the answer is distressingly simple: with a username and a password, often hardcoded directly into the bot's script or stored in a loosely secured configuration file. These credentials, meant for a single, automated process, effectively become a static, long-lived key to one or more critical applications. If an attacker can find this script or access the file, a task made easier if it is stored in a poorly secured code repository or network share, they have everything they need to impersonate the bot.

This leads to the second major risk: the principle of least privilege is almost universally violated for bots. A human employee assigned to process invoices might only have permission to enter data into the accounting system. A bot designed for the same task, however, is often given a highly privileged generic service account with broad access, simply because it is easier than scoping its permissions precisely. The bot might have the ability not just to enter invoices but to approve payments, modify vendor records, and run financial reports. An attacker who compromises the bot's credentials doesn't just gain the ability to automate a task; they inherit a treasure trove of excessive permissions, providing a powerful launchpad for lateral movement and fraud.

The problem is compounded by the bot orchestration platforms themselves. These are the centralized consoles used to manage the bot workforce, schedule their tasks, and store their credentials. The orchestrator is the bot's boss, and compromising it is the equivalent of taking control of the entire automated division of the company. An attacker with access to the orchestration platform could subtly alter the logic of existing bots to exfiltrate data, or they could create new, malicious bots and deploy them into the environment, all under the guise of legitimate business automation. The very tool designed for efficiency becomes a weapon of mass compromise.

While bots mimic human users, another, even larger class of non-human identity operates entirely behind the scenes, forming the hidden connective tissue of the modern digital world. This is the world of the Application Programming Interface, or API. Today's applications are rarely built as single, monolithic blocks of code. They are constellations of smaller, specialized microservices that communicate with each other and with third-party platforms through APIs. Each of these services needs an identity and a way to authenticate itself before it can be authorized to request data or trigger an action. This machine-to-machine communication is the silent, humming engine of the digital economy, and it vastly outnumbers human-led interactions.

The identities of these services are typically managed through credentials like API keys, OAuth 2.0 access tokens, or mutual TLS certificates. These credentials, like the hardcoded passwords of RPA bots, are a prime target for attackers. The internet is littered with stories of developers accidentally committing a secret API key to a public GitHub repository. Automated scanners operated by adversaries find these leaked keys within minutes, giving the attacker a direct, authenticated channel into a company's backend systems.

This problem is exacerbated by the phenomenon of "shadow APIs" and "zombie APIs." As applications evolve, developers create new API endpoints and refactor old ones. In the rush to innovate, they often forget to properly decommission the older versions. These forgotten endpoints, or zombie APIs, may still be active and connected to production databases but are no longer documented, monitored, or maintained. A shadow API is one that was created for a temporary or internal purpose and never formally documented at all. An attacker who discovers one of these forgotten backdoors has found a perfect point of entry, one that is likely unmonitored and may have lax security controls.

Even when an API is properly documented and secured, it can be vulnerable to flaws in its authorization logic. One of the most common and dangerous vulnerabilities in the API world is known as Broken Object Level Authorization (BOLA). This occurs when an API endpoint fails to properly verify that the user making a request has the right to access the specific resource they are asking for. For example, an API call to retrieve a user's profile might look like this: `GET /api/v1/users/123`. The system correctly validates the authentication token of the logged-in user. However, if that user simply changes the number in the URL to `/api/v1/users/456` and the system returns the profile of a different user without checking if the requester is authorized to see it, that is a BOLA vulnerability. It allows any authenticated user to potentially cycle through every user ID and exfiltrate the entire user database.

A similar problem exists in the world of federated APIs that use the OAuth 2.0 protocol. When a third-party application wants to access a user's data, it requests a specific set of permissions, or "scopes." Users are often conditioned to quickly click "approve" on these consent screens without carefully reading what they are agreeing to. A mobile game might legitimately need access to a user's basic profile, but it might also request the scope to read all of their contacts. If the user approves this, and that gaming company is later breached, the attackers gain access to not just the user's game data but their entire contact list. The overly permissive scope, granted once and then forgotten, becomes a persistent liability.

The third and perhaps most unsettling emerging attack surface is not about compromising an existing identity, human or machine. It is about creating a trustworthy identity from nothing. This is the world of synthetic identity fraud, a sophisticated form of deception that has been supercharged by the advent of artificial intelligence. A synthetic identity is not a stolen identity. It is a fabricated persona, a carefully constructed mosaic of real and fictitious data designed to create a person who does not exist but who appears entirely plausible to an automated system.

The process often begins with a real, but stolen, Social Security Number, typically one belonging to a child, a deceased person, or someone who is unlikely to have an existing credit history. This real piece of data is then combined with a fabricated name, address, and date of birth. The attacker then begins to patiently build a life for this synthetic person. They might use the identity to apply for a credit card, get rejected, and then try again in a few months. Eventually, this activity creates a thin, but real, credit file for the fabricated identity. They might use generative AI to create a realistic profile picture, a plausible employment history on LinkedIn, and a smattering of social media activity.

Over time, this synthetic identity becomes more and more "real" in the eyes of the digital world. It has a credit history. It has an online presence. It looks and feels like a real person. This poses a fundamental challenge to traditional Identity Verification (IDV) processes. Most IDV systems are designed to spot inconsistencies in a real person's data,does the name on the driver's license match the name in the credit bureau's database? A synthetic identity, having been built from the ground up to be internally consistent, often sails through these checks.

The attack scenarios for these synthetic identities in an IAM context are chilling. An adversary could use a carefully cultivated synthetic identity to apply for a job at a target company. With a believable resume and a clean, albeit fabricated, background, they might pass the automated HR screening and even a cursory human review. Once hired, this synthetic employee is a malicious insider from day one, with legitimate credentials, a corporate-issued laptop, and authorized access to internal systems. They can operate for months, mapping the network and exfiltrating data, before disappearing without a trace. The person the company is trying to fire never actually existed.

The same technique can be used to onboard synthetic customers or partners. An attacker could create thousands of synthetic customer accounts to abuse sign-up bonuses, overwhelm an application with traffic, or write floods of fake reviews. They could establish a synthetic company, complete with a professional-looking website and fabricated employee profiles, and apply to become a trusted vendor in a corporation's supply chain. Once accepted, they have a trusted position from which to launch more sophisticated attacks, such as submitting malicious invoices or gaining access to shared partner portals.

Governing this new, diverse population of non-human and fabricated identities with our old, human-centric tools is a losing battle. The traditional access certification campaign, where a manager reviews their direct reports' permissions, is meaningless for a bot or an API key that has no manager. Role-Based Access Control struggles to define a "role" for a microservice that has a very specific and narrow set of tasks that may not align with any human job function. The identity lifecycle management processes of "joiner, mover, leaver" do not apply to a synthetic identity that was never truly a joiner in the first place.

This new reality demands a new approach to governance, one that treats machine identity as a first-class discipline. It starts with a comprehensive discovery and inventory process, using specialized tools to find every bot, service account, and API key in the environment, not just the ones that are officially documented. Each of these identities must have a clearly defined owner, a purpose, and a lifecycle. Their credentials should not be static and long-lived but managed through automated secret vaults that issue short-lived, just-in-time credentials that are automatically rotated. Their permissions must be scoped with ruthless precision, based on an analysis of their actual behavior, not on a convenient, overly permissive template.

For bots and APIs, behavioral analytics is even more effective than it is for humans. Machines are supposed to be predictable. A service account that has only ever performed read operations on a single database table and suddenly attempts to delete the entire database is a five-alarm fire. The baseline for a machine is rigid, and any deviation is highly suspect.

Combating synthetic identities requires moving beyond simple document and data verification. It requires a more holistic, network-level analysis. Sophisticated anti-fraud systems look for subtle, statistical patterns that suggest fabrication. Does a single device seem to be applying for multiple new accounts? Are multiple applicants using the same phone number or address? Is the data in the application, while internally consistent, too perfect or too generic? It is a shift from verifying an identity to verifying its "humanness" and its plausible existence in the real world.

Ultimately, the emergence of these new attack surfaces forces us to broaden our very definition of identity. The entities we must manage are no longer just the people on our payroll. They are the automated scripts in our data centers, the API-driven services that connect our applications, and the fabricated ghosts that knock on our digital doors. Each one requires a unique approach to authentication, authorization, and governance. Without expanding our vision to include this new and growing population, we are leaving our most critical digital doors wide open, watched over by guards who have been trained only to recognize a human face.

* * *

23

Navigating Regulation: Global Compliance and AI Governance

The life of an Identity and Access Management architect used to be relatively straightforward, at least from a legal perspective. The primary concerns were internal corporate policies and a handful of well-understood industry standards. The work was technical, logical, and largely confined to the digital realm. That world is gone. Today's IAM leader must be as much a diplomat and amateur legal scholar as they are a technologist. The systems they build are no longer judged solely on their security and efficiency; they are now subject to a complex and rapidly evolving patchwork of global regulations governing everything from data privacy to the ethical framework of artificial intelligence itself. Navigating this new landscape is not a matter of ticking boxes on a compliance checklist. It is a dynamic challenge of architectural design, requiring systems that are not just intelligent, but also worldly, multi-lingual in the language of law, and acutely aware of the digital borders they operate across.

The compliance burden has traditionally been viewed through the lens of data privacy, with regulations like the EU's GDPR setting the gold standard. These laws established foundational principles like data residency, user consent, and the rights of data subjects. While immensely important, this is now just the first layer of a much more complex regulatory cake. The modern IAM architect must contend with a new trinity of compliance concerns that intersect and often conflict, creating a formidable governance puzzle. The first pillar remains data privacy, but the challenge has morphed from understanding one or two major laws to managing a mosaic of dozens of national and regional variants, from California's CPRA to Brazil's LGPD and India's DPDP Act, each with its own nuances.

The second, and most transformative, pillar is the rise of AI-specific regulation. This is a new breed of law that moves beyond governing data to governing the algorithms that process it. The landmark example is the European Union's AI Act, a comprehensive framework that classifies AI systems based on their potential risk to society. An AI system used to recommend music is low-risk. An AI system that makes the final decision on whether to grant a person a mortgage or access to a critical government service is deemed "high-risk." Given that modern IAM platforms increasingly use AI to make autonomous or semi-autonomous decisions about a person's ability to work, bank, and access services, it is almost certain that many IAM 3.0 systems will fall into this high-risk category.

This classification is not merely a label; it triggers a cascade of stringent legal obligations. A high-risk AI system under the EU AI Act must undergo a rigorous conformity assessment before it can be deployed, similar to getting a medical device approved. It must be built on a foundation of high-quality, representative training data to minimize the risk of bias. It must have a robust risk management system in place for its entire lifecycle. Crucially, it must be designed to allow for effective human oversight, ensuring that a person can intervene or override a machine's decision. And it must meet high standards of transparency, providing clear information to users and generating logs that are detailed enough to be audited. These are not IT best practices; they are hard legal requirements, and failure to comply can result in fines that make even the steepest GDPR penalties look modest.

This same risk-based approach is now being replicated in other jurisdictions, from Canada to the United Kingdom, creating a new global baseline for AI governance.

The third pillar of this new compliance reality is the thicket of sector-specific mandates that intersect with both privacy and AI. The rules have always been different for a bank, a hospital, or a power plant. Now, these existing regulations are being reinterpreted for the age of intelligent identity. In financial services, regulators are scrutinizing automated systems to ensure they do not perpetuate discriminatory lending practices, a direct challenge to potentially biased AI risk models. In healthcare, the Health Insurance Portability and Accountability Act (HIPAA) in the US sets strict rules on the use of patient data. Using an AI to analyze clinician behavior to grant access to medical records must be done in a way that is fully compliant with these long-standing patient privacy rules. For critical infrastructure, governments are imposing new cybersecurity requirements that increasingly dictate how access is managed, often with specific controls that may not align with a fluid, probabilistic IAM model. The IAM architect must now layer these industry-specific constraints on top of the broader privacy and AI regulations, creating a multi-dimensional compliance matrix.

To build a system that can survive this regulatory crossfire, the architecture itself must be designed with compliance as a core, non-functional require-ment. The old model of a single, global IAM platform is no longer tenable. It must be replaced by a federated architecture of sovereign instances, or "pods," as introduced in Chapter Seventeen. Each pod is a self-contained deployment of the IAM stack, including the AI models and the data they process, that resides within a specific legal jurisdiction. A company might have an EU pod running in a Frankfurt data center, a North American pod in Virginia, and an APAC pod in Singapore.

The challenge then becomes one of managing policy across this distributed fleet. The solution is to separate the plane of policy definition from the plane of policy enforcement. A central team of architects and governance specialists can define a global baseline of access policies. However, these policies are not monolithic. They are built with conditional logic that references the jurisdiction. For example, a global policy might be written in a human-readable language like YAML or JSON, and then compiled and pushed out to the local pods.

```
<pre><code>
  - policy_name: 'Admin_Access_Risk_Threshold'
  description: 'Sets the maximum allowable risk score for granting admin access.'
  global_default:
  risk_score_limit: 0.2
  overrides:
  - jurisdiction: 'EU'
  risk_score_limit: 0.15
  justification: 'Higher standard required by internal interpretation of EU AI Act for high-risk systems.'
  - jurisdiction: 'CN'
  risk_score_limit: 0.1
  justification: 'Stricter local data security regulations.'
  </code></pre>
```

This "regulation-as-code" approach makes compliance logic explicit, version-controlled, and auditable. It provides a scalable way to manage a complex global policy landscape while allowing for the necessary local variations mandated by law. The German pod enforces the German rules, the Chinese pod enforces the Chinese rules, and the central governance team maintains a unified view of the overall policy framework.

This new architecture also demands a new kind of data pipeline dedicated specifically to generating compliance evidence. In the past, audit evidence was often gathered in an ad-hoc manner, with teams scrambling to pull logs and reports in response to a specific request. This is no longer sufficient. An organization must be able to prove, on demand, that its AI systems are operating fairly and safely. This requires an automated pipeline designed to produce auditable artifacts.

The inputs to this pipeline are the outputs of the IAM system, particularly the explanations generated by the XAI layer. Every time an AI makes a high-stakes decision,like denying access or locking an account,the decision and its detailed, human-readable justification are fed into the compliance pipeline. This pipeline aggregates the data and produces a series of "model health" reports and dashboards. These are not for real-time security monitoring but for long-term governance. One dashboard might track the rate at which the model's decisions are appealed by users, with a breakdown by user demographic to spot potential bias. Another might track the model's performance against a known set of test cases to detect model drift over time. This creates a continuous, evidence-based record of the model's behavior, ready for an internal or external audit.

The ultimate output of this pipeline can be a "model card" or a "decision audit report," a standardized document that summarizes a model's characteristics. Much like a nutrition label on a food package, a model card provides essential information in a clear format: the model's purpose, its performance metrics, the characteristics of the data it was trained on, the results of bias testing, and any known limitations. This proactive generation of evidence shifts the posture from reactive scrambling to "continuous compliance," where proof of control is a standard output of the system's operation.

Of course, technology and architecture are only part of the solution. The human governance structures surrounding the AI must also evolve. The responsibility for an AI's behavior cannot rest solely on the shoulders of the CISO or a single data scientist. It is a shared, cross-functional responsibility. This has led to the emergence of formal AI Ethics Boards or AI Risk Committees within forward-thinking organizations. This is not a technical review board; it is a governance body composed of senior leaders from legal, privacy, human resources, data science, information security, and the relevant lines of business.

This committee's charter is to provide the human oversight mandated by the new regulations. Their duties include reviewing and formally approving any new high-risk AI model before it is deployed into production. They are responsible for setting the organization's risk appetite, defining what constitutes an acceptable level of bias or error. They oversee the appeals process, acting as the ultimate court of appeal for users who believe they have been wronged by a machine's decision. And they are responsible for reviewing the continuous monitoring reports, watching for model drift or the emergence of new biases over time. This provides a crucial layer of human accountability, ensuring that the organization's use of AI remains aligned with its legal obligations and ethical principles.

This new world also requires new skills. The neat lines between professions are blurring. An IAM architect who cannot read and understand the basic tenets of the EU AI Act is a liability. A lawyer providing advice on AI risk who does not understand the difference between model training and inference is ineffective. A data scientist who builds a brilliant risk model without considering its potential for discriminatory impact is creating a future legal crisis. Organizations must invest in cross-training, creating a shared language and a shared understanding of the issues that sit at this complex intersection of law, data science, and security.

One of the most direct and challenging new legal concepts to operationalize is the "right to explanation." Several new regulations, including GDPR, allude to a user's right to receive a meaningful explanation for an automated decision that significantly affects them. Implementing this requires more than just having an XAI tool. It requires a formal, user-facing process. An organization must be able to receive a user's request for an explanation, route it to the correct team, use its XAI capabilities to generate a justification that is both faithful to the model's logic and understandable to a non-technical person, and deliver it to the user within a legally mandated timeframe. This entire workflow must be designed, staffed, and managed.

This is all leading to a new era of third-party certification and auditing for AI systems. Just as organizations undergo SOC 2 or ISO 27001 audits for their security controls, they will soon face audits for their AI governance programs. Standards bodies are already developing frameworks for AI management systems. An auditor of the future will not just ask for firewall rules and password policies. They will ask to see your AI model inventory. They will demand the model cards for your high-risk systems. They will want to review the minutes of your AI Ethics Board meetings and inspect the results of your bias testing. They will scrutinize the design of your appeals process. Preparing for this new world of AI assurance requires building the necessary evidence-gathering and reporting capabilities into your IAM systems from day one.

Finally, architects must grapple with the practical challenge of managing cross-border data flows in this new, fragmented world. The classic approach of using legal agreements like Standard Contractual Clauses (SCCs) to justify international data transfers is coming under increasing strain, especially when the "transfer" is an ephemeral, machine-to-machine API call between two AI services. This has spurred immense interest in a class of technologies known as Privacy-Enhancing Technologies, or PETs. These are advanced cryptographic and statistical methods that allow for data to be analyzed without being exposed.

For example, a technique called federated learning could allow a global company to train a single, unified behavioral model without ever moving raw user data out of its sovereign pod. The central system sends the model structure to each local pod. The model is then trained locally on the sovereign data. Only the resulting changes to the model's parameters,a series of abstract mathematical weights, not the user data itself,are sent back to be aggregated into the global model. Other techniques, like homomorphic encryption, could one day allow a risk engine to perform calculations directly on encrypted data without ever having to decrypt it. While many PETs are still emerging from the laboratory, they represent a potential future where we can achieve the benefits of global intelligence without breaking the laws of digital sovereignty, offering a technical solution to what has become a deeply political problem.

* * *

24

Building a Resilient, Future-Ready IAM

The journey through the landscape of IAM 3.0 can feel like a tour of an arsenal where every new weapon of defense is displayed next to an equally powerful and often more subtle weapon of attack. We have explored the transformative intelligence of AI and the probabilistic reasoning it enables, only to be confronted by the specter of adversarial machine learning. We have seen the promise of a seamless, federated world, only to recognize it as a superhighway for compromise. We have contemplated a future of user-controlled, decentralized identity, only to run into the paradox of the all-seeing risk engine. And looming over it all is the quantum time bomb, silently threatening to invalidate the very mathematics of trust.

This confluence of opportunity and peril can lead to a state of paralysis. Faced with such complex and interwoven challenges, the temptation is to retreat to the familiar, to double down on old controls and hope for the best. But the threats we face are not standing still. A future-ready identity program cannot be built on a foundation of nostalgia or fear. It must be built on a clear-eyed acceptance of this new reality and a strategic commitment to a new set of architectural principles. The goal is no longer to build an impenetrable fortress; the perimeter is gone, and the adversary is already inside. The new objective is to build a resilient ecosystem,one that can absorb shocks, adapt to change, and continue to function securely in a state of constant contention.

A resilient architecture begins with a fundamental shift in mindset: from rigidity to agility. For decades, security was built on the principle of creating strong, static defenses. The goal was to configure the firewall, harden the server, and write the access policy, and then defend that configuration against all change. This approach is too brittle for the modern environment. An agile, resilient architecture assumes that everything is in a constant state of flux: the users, the applications, the networks, the threats, and even the cryptographic algorithms themselves. The system must be designed to accommodate this change as a normal state of operation, not as an exception.

This leads directly to the principle of modularity. A monolithic IAM platform, where all components are tightly coupled, is a significant liability. A flaw in one part can bring down the entire system, and upgrading any single component can become a massive, high-risk project. A future-ready architecture is composed of smaller, independent, and loosely coupled services that communicate through well-defined APIs. Your identity governance engine should be distinct from your adaptive authentication engine, which in turn should be separate from your credential management service. This modularity allows you to upgrade, replace, or reconfigure individual components without destabilizing the entire ecosystem. It allows you to introduce a new PQC-compliant signing service, for example, by simply redirecting an API call, rather than re-architecting your entire identity provider.

This architectural agility must be supported by a disciplined practice of treating your infrastructure and policies as code. The configuration of your IAM system,the access policies, the risk thresholds, the trusted issuer lists,should not live in a graphical user interface. It should be defined in human-readable, machine-enforceable text files stored in a version control system like Git. This practice of "Identity-as-Code" has profound benefits for resilience. It makes every change to your security posture transparent, auditable, and repeatable. If a flawed policy is deployed, you can instantly roll back to a known-good version. You can create automated testing pipelines that validate your policies for correctness and security before they ever reach production. It transforms the art of IAM configuration from a manual, error-prone craft into a rigorous, reliable engineering discipline.

Of course, the brain of this new ecosystem is its intelligence layer, and the resilience of the entire system hinges on the trustworthiness of that brain. As we have seen, deploying a powerful AI model without the ability to understand its reasoning is the equivalent of hiring a security guard who speaks a language no one else understands. Therefore, a non-negotiable principle for a future-ready IAM program is that intelligence must be coupled with introspection. Explainable AI (XAI) and Access Observability are not optional, premium features; they are the essential safety mechanisms that make intelligent IAM viable.

This principle must be operationalized, not just discussed. It begins with procurement. Your Request for Proposal (RFP) for a new IAM vendor must include a detailed section on explainability. It should demand that the vendor demonstrate not just the accuracy of their AI but its ability to justify its decisions in a format tailored for different audiences: security analysts, auditors, and end-users. The same rigor must be applied to observability. A vendor must prove that their system can export the rich, structured, and correlated telemetry needed to trace a single access decision from start to finish. A platform that cannot be explained or observed is a black box that has no place in a resilient security architecture.

Once deployed, this intelligent infrastructure must be treated as its own potential attack surface. We spend vast sums on penetration testing for our applications and networks, but very few organizations have a program for actively trying to break their own AI models. Building a resilient program requires establishing a dedicated "AI Red Team" or an adversarial testing function. The job of this team is to think like the adversary and actively attack your models. They should be running automated probes to find evasion loopholes. They should be designing and simulating data poisoning attacks to see if they can manipulate your behavioral baselines. They should be testing the XAI layer to see if it can be tricked into producing a misleading explanation. This proactive, adversarial mindset is the only way to find the weaknesses in your intelligent systems before a real attacker does.

This rigorous governance must extend to the rapidly growing population of non-human identities. A resilient program cannot have a blind spot that encompasses the majority of its active entities. The practice of Machine Identity Management must be elevated to a first-class discipline, on par with human identity governance. This starts with a unified, automated discovery and inventory process to find every bot, service, and API key in the environment. Each of these identities must have an owner, and its entire lifecycle,from creation to rotation to retirement,must be automated. Static, long-lived credentials like passwords and API keys should be aggressively eliminated and replaced with short-lived, just-in-time certificates and tokens issued from a centralized secrets vault. This drastically shrinks the window of opportunity for an attacker to misuse a compromised machine credential.

The behavior of these machine identities is a rich source of security intelligence. As machines are meant to be predictable, any deviation from their established baseline is a high-fidelity signal of potential compromise. Your access observability platform must ingest the activity logs of these non-human entities, and your behavioral analytics models must be trained to recognize their unique patterns. The goal is to apply the same level of continuous, risk-based scrutiny to a service account trying to access a database that you apply to a human user logging into a laptop.

Ultimately, resilience is a socio-technical property. The most brilliantly designed architecture can be undone by a team that lacks the skills to operate it or a culture that resists change. Building a future-ready program, therefore, is as much about investing in people and processes as it is about investing in technology. Your security operations team needs a new set of skills. Analysts who were trained to read raw syslog data must now be trained to interpret the output of an XAI engine and navigate a complex access observability platform. They need to evolve from log jockeys to data detectives.

This requires a new level of collaboration. The walls between the IAM team, the security operations center, the data science group, and the GRC team must come down. They must work from a common playbook, using the access observability platform as their shared source of truth. When a potential model bias is detected, it should trigger a collaborative investigation involving all these stakeholders. This cross-functional approach is essential for governing the complex trade-offs between security, fairness, and usability that are inherent in an AI-driven system.

Realizing this vision of a resilient, future-ready IAM program can seem overwhelming. It is not a single project but a continuous journey of maturation. The most effective way to approach it is with a phased, multi-year strategy that prioritizes foundational capabilities first and builds toward more advanced ones over time.

The first phase, which should be the immediate focus for any organization, is about establishing foundational visibility and hardening the core. This means aggressively pursuing a comprehensive inventory of all cryptographic assets and all machine identities. You cannot manage what you do not know you have. Concurrently, you must begin the work of building out your access observability platform, instrumenting your critical systems to provide the necessary telemetry. The other immediate priority is to attack the greatest single source of risk: passwords. A concerted push to deploy strong, phishing-resistant passwordless authentication, like FIDO2, for all users provides the biggest security return on investment in the short term.

The second phase is about embracing intelligence with caution and control.

This is the time to begin piloting the use of AI and machine learning for adaptive access, but only with the non-negotiable prerequisite that the chosen platform provides robust and usable explainability. During this phase, you should formalize the creation of your AI governance body and begin building the internal muscle for adversarial testing of your models. This is also the time to get serious about planning for the quantum future by engaging with your key vendors on their PQC roadmaps and starting to identify your most at-risk, long-lived data.

The third and final phase is about future-proofing and cautiously moving toward automation. With a solid foundation of observability and explainable AI in place, you can begin to pilot more advanced capabilities. This is the time to experiment with Trust Mediation Engines to handle interactions with the emerging decentralized identity ecosystem. It is also when you can begin carefully piloting autonomous identity features, but only for well-understood, low-risk use cases, such as managing access for temporary contractors in a sandboxed environment. Any move toward autonomy must be accompanied by intense monitoring and a clear human override and appeals process.

This journey is not easy, and it is never truly finished. The threat landscape will continue to evolve, new technologies will emerge, and the regulatory environment will shift again. A resilient IAM program is not a destination; it is a state of being. It is an organizational commitment to continuous learning, adaptation, and introspection. It requires us to move beyond the search for a perfect, static defense and instead embrace the design of dynamic, observable, and intelligent systems that can withstand the inevitable pressures of a complex and uncertain world.

* * *

25

The Next Frontier: Human, Machine, and Synthetic Identity Trust

The history of digital identity has been a story of progressive abstraction. We moved from managing accounts on a single computer to managing users within a network, and then to managing federated identities across a global ecosystem. At each stage, the core subject remained comfortably constant: the human being. The entire edifice of access control,the policies, the psychology of passwords, the user experience design,was built around the concept of a person sitting at a device. The IAM 3.0 paradigm, with its focus on behavior and context, represents the pinnacle of this human-centric model. Yet, as we stand on this peak, we can see the terrain shifting once again, revealing a new and far more complex frontier.

This new landscape is no longer a simple bimodal world populated by people and a few well-behaved service accounts. It is a teeming, chaotic ecosystem of entities, a digital menagerie where the lines between human, machine, and fabricated personas are becoming increasingly, and intentionally, blurred. Our next great challenge is not just to secure the human user, but to build a coherent framework of trust that can accommodate this new, diverse, and often deceptive population. The next frontier of identity is not about a single entity type; it is about establishing a universal grammar of trust that can be spoken by all of them, whether they are made of flesh and blood, silicon and logic, or pure information.

The human identity, for all the focus it has received, is still an evolving concept. We have made tremendous strides in moving beyond the fragile password, embracing strong, phishing-resistant authenticators and continuously evaluating behavioral signals. The frontier of human trust, however, is pushing into even more nuanced territory. It is about understanding not just the what of a user's actions, but the why. This involves moving from coarse-grained signals like location and device to finer-grained indicators of intent and context.

Imagine an IAM system integrated with an organization's collaboration tools. The system might observe that a user has just been added to a high-urgency chat channel for a critical incident involving a specific database. When that same user almost immediately attempts to request just-in-time access to that very database, the system has a powerful contextual signal that this is a legitimate, mission-critical action. The trust score is not just based on the user's baseline behavior but on the semantic context of their work at that precise moment. This requires a level of data integration and natural language understanding that we are only just beginning to explore, but it points to a future where trust assessment is deeply woven into the fabric of daily work.

This journey also brings with it profound ethical quandaries. As we seek more fidelity in our trust signals, it is a short and tempting step from observing behavior to inferring a user's emotional or cognitive state. Could a system analyze keystroke patterns to detect stress or distraction? Could it use sentiment analysis on communications to flag a "disgruntled employee" as a higher risk? The technical ability to do so is not far off, but the ethical and privacy implications are staggering. The frontier of human trust will require us to draw very clear and very firm lines, defining what data is acceptable for a machine to use in judging a person and codifying those ethical boundaries into our systems as hard policy.

At the same time, we must contend with the fact that the human user is no longer acting alone. They will be increasingly augmented by their own personal AI agents,digital assistants with delegated authority to manage their calendar, answer their emails, and execute tasks on their behalf. When an AI agent, acting on behalf of a human, requests access to a file, whose identity is it? Who is accountable? This creates the need for a model of delegated identity, where a user can grant a specific, scoped, and time-bound set of permissions to their own AI agent. The trust model must be able to distinguish between the human principal and their machine agent, applying different rules and scrutiny to each, even when they are acting in concert.

The second great population of our new ecosystem is that of the machines. The sheer volume of machine identities,from microservices and APIs to RPA bots and IoT devices,has already dwarfed the human population, yet our models for managing their trust remain primitive. We have moved beyond hardcoding passwords in scripts, but we are still largely reliant on static credentials like API keys and OAuth tokens managed in vaults. The next frontier of machine trust is to treat these entities with the same behavioral sophistication we apply to humans. It is about giving each machine a "digital DNA," a baseline so unique that any deviation is a clear signal of compromise.

For a microservice, this digital DNA would be a multi-faceted profile. It would include its normal network communication patterns: which other services does it talk to? Over which ports? How much data does it typically send and receive? It would include a baseline of its resource consumption on the host server: its typical CPU and memory usage, and the system calls it normally makes. It would even include a profile of the code itself. Using techniques like remote attestation, a system could continuously verify that the cryptographic hash of the service's running code matches the official version in the code repository. Any unauthorized change to the code, however small, would invalidate its identity.

When a machine identity is assessed, the risk engine would compare its real-time activity against this multi-dimensional baseline. A service that suddenly tries to connect to an unknown external IP address, whose memory usage spikes unexpectedly, or whose code signature no longer matches the approved version, would have its trust score plummet. This allows us to move from a reactive model of rotating a secret after it's been leaked to a proactive model of detecting a compromised machine based on its aberrant behavior.

This approach becomes even more critical for the growing army of RPA bots. A bot's behavior should be brutally predictable. It is programmed to follow a specific workflow, clicking on the same buttons in the same sequence, day after day. A behavioral analytics engine trained on this workflow could detect the slightest deviation. If a bot that normally only ever reads data from a CRM suddenly attempts to use the "export all records" function, it is an almost certain indicator that its session has been hijacked or its logic maliciously altered. By treating the bot not as a person but as a highly constrained automaton, we can hold it to a much stricter standard of behavioral normalcy.

The governance of these machine identities must also mature. The concept of an "owner" must be strictly enforced for every non-human identity. Every service account, every bot, every API key needs to be tied to a human being or a team who is responsible for its existence. This creates a clear line of accountability. The lifecycle of these identities must be aggressive and automated. A credential for a temporary data processing job should be created just-in-time and should cease to exist the moment the job is complete. The default state for a machine identity should be non-existence, with access being granted only for the fleeting moments it is required.

The third, and most challenging, population on this new frontier is the one that shouldn't exist at all: the synthetics. These fabricated personas, as we have seen, are designed to perfectly mimic the data trail of a real human, making them exceptionally difficult to spot with traditional identity verification tools. Combating this threat requires a fundamental shift in our thinking. We must move beyond simply verifying an identity,checking that the presented data is consistent,to actively affirming its plausible existence in the physical world. This is a much harder, more investigative problem.

The frontier of synthetic identity detection is about finding the subtle statistical artifacts of fabrication. It is a form of digital forensics that looks for the tell-tale signs of a constructed life. Advanced fraud platforms are no longer just checking a name against a credit file. They are performing complex, network-level analysis. They look for impossibly clean data; real people have typos, old addresses, and messy data histories, while synthetics are often too perfect. They look for correlations that are statistically improbable. Do dozens of "different" new account applications originate from the same device or a small block of IP addresses? Does a single phone number appear to be associated with multiple, otherwise unrelated identities? These network graph analysis techniques can reveal the hidden puppeteer behind a crowd of seemingly independent synthetic puppets.

Another emerging technique is to analyze the "digital exhaust" of an identity. A real person, over years of online activity, creates a rich, complex, and often contradictory trail of data across hundreds of different services. A synthetic identity, by contrast, often has a very thin, curated, and recent history. Affirmation systems can perform deep web searches to look for this corroborating evidence. Does this person have a social media presence that looks organic and has existed for years? Do they appear in public records, news articles, or other third-party data sources in a way that aligns with their claimed history? The absence of this rich, messy digital footprint can itself be a strong signal of fabrication.

The ultimate challenge, then, is to build an IAM framework that can manage and apply these three distinct trust models,nuanced behavioral analysis for humans, rigid baselining for machines, and investigative affirmation for synthetics,under a single, coherent governance umbrella. A single access request could now originate from any of these entity types, and the system must be able to first classify the entity and then apply the appropriate trust calculus.

This requires an architecture that is more abstract and signal-driven than ever before. At the heart of this future system might be a "Universal Trust Engine." This engine would not have a monolithic model but a library of specialized risk assessment modules. When a request comes in, a preliminary classification service would attempt to determine the entity type. Is this a known human employee? A registered service account? A brand new customer from an unknown source?

Based on this classification, the request would be routed to the appropriate module. A request from a known employee would be sent to the human behavioral analytics and XAI module. A request from a registered microservice would be routed to the machine DNA validation module. A new customer onboarding request would be sent to the synthetic identity affirmation module. Each module would perform its specialized analysis and produce a standardized set of output signals: a confidence score, a set of risk factors, and an evidence package.

These signals would then be fed into a final policy orchestration layer. This layer would consume the standardized outputs from the various modules and make the final access decision based on a unified policy framework. A policy could be written that says, "To access the customer payment system, the requesting entity must have a confidence score above 0.9, regardless of whether it is a human finance clerk, an automated billing bot, or a partner API." This allows for consistent governance while accommodating the very different ways that trust is assessed for each entity type.

This future vision brings the IAM 3.0 journey full circle. We began by breaking the old paradigm of static, deterministic rules. We introduced the intelligence to understand the subtle context of human behavior. Now, we must expand that intelligence to encompass a world where identity itself has become plural. The next great task is to build a system that is not only smart enough to tell friend from foe, but wise enough to understand that its "friends" and "foes" may come in forms we are only just beginning to imagine. It is about creating a system of trust that is resilient enough not just for the identities we have today, but for all the identities, human and otherwise, that are yet to come.

* * *

www.ingramcontent.com/pod-product-compliance
Lightning Source LLC
Chambersburg PA
CBHW061312220326
41599CB00026B/4841